New Frontiers in Organoselenium Compounds

AF822775

Eder João Lenardão • Claudio Santi
Luca Sancineto

New Frontiers in Organoselenium Compounds

Springer

Eder João Lenardão
CCQFA - LASOL
Universidade Federal de Pelotas
Pelotas, Rio Grande do Sul, Brazil

Claudio Santi
Department of Pharmaceutical Sciences
Università degli Studi di Perugia
Perugia, Perugia, Italy

Luca Sancineto
Section of Heterorganic Chemistry
Centre of Molecular and Macromolecular
Studies, Polish Academy of Sciences
Łódź, Poland

ISBN 978-3-030-06426-6 ISBN 978-3-319-92405-2 (eBook)
https://doi.org/10.1007/978-3-319-92405-2

© Springer International Publishing AG, part of Springer Nature 2018
Softcover re-print of the Hardcover 1st edition 2018
This work is subject to copyright. All rights are reserved by the Publisher, whether the whole or part of the material is concerned, specifically the rights of translation, reprinting, reuse of illustrations, recitation, broadcasting, reproduction on microfilms or in any other physical way, and transmission or information storage and retrieval, electronic adaptation, computer software, or by similar or dissimilar methodology now known or hereafter developed.
The use of general descriptive names, registered names, trademarks, service marks, etc. in this publication does not imply, even in the absence of a specific statement, that such names are exempt from the relevant protective laws and regulations and therefore free for general use.
The publisher, the authors and the editors are safe to assume that the advice and information in this book are believed to be true and accurate at the date of publication. Neither the publisher nor the authors or the editors give a warranty, express or implied, with respect to the material contained herein or for any errors or omissions that may have been made. The publisher remains neutral with regard to jurisdictional claims in published maps and institutional affiliations.

Printed on acid-free paper

This Springer imprint is published by the registered company Springer International Publishing AG part of Springer Nature.
The registered company address is: Gewerbestrasse 11, 6330 Cham, Switzerland

Foreword

The book *New frontiers on organoselenium compounds*, by Lenardão, Santi, and Sancineto, is destined to become an extremely useful and valuable source for students and researchers in the field. It is also an excellent tribute to the many researchers who were involved in the field of organochalcogen chemistry. My first impression upon reading this book was that it is uniformly elegant and organized logically. This book has numerous vignettes on the chemistry of organochalcogens and their biological aspects in a very appealing way. The context of each topic is made crystal clear by the inclusion of appropriate figures/schemes, and the references are eclectic as well as up to date.

Through the four chapters, written by renowned experts in the field, the contents are well organized with an incredible amount of results without exclusion of important topics.

In the first chapter, the application of organoselenium compounds as reagents and catalysts in organic synthesis with a focus on the development of sustainable methods is discussed. The author organized this chapter comprehensively with updated literature.

The biologically relevant organoselenium compounds and their therapeutic prospects as well as their historical background are discussed in the second chapter. An overview on the most active organoselenium small molecules with an emphasis on the GPx-like activity is described. There is also discussion on antiviral, antibacterial, antifungal, and antiprotozoal activities.

"Organoselenium in nature," which is the third chapter, describes the role of naturally occurring selenium compounds in humans, plants, and bacteria. The importance of selenocysteine (L-Sec) and selenoenzymes and their molecular mechanism are discussed.

The last chapter is devoted to the non-bonded interaction of chalcogens in organochalcogen compounds. These interactions are discussed under the perspective of proteomics as well as in drug discovery processes. This chapter also highlights some of the recent reports involving these interactions in material chemistry.

New frontiers on organoselenium compounds was structured for multidisciplinary areas by putting together advancement ranging from organic to biochemistry, synthesis to structural chemistry. I highly recommend this book for academia as well as for industry.

Federal University of Santa Catarina – UFSC
Florianopolis, Brazil
April 12, 2018

Antonio L. Braga

Contents

Chapter 1
Organoselenium Compounds as Reagents and Catalysts to Develop New Green Protocols

Abstract This chapter offers a comprehensive and updated overview on the use of organoselenium compounds as catalysts in organic synthesis. Around 150 references were carefully revised, covering the recent advances in established Se-catalyzed reactions, like the hydrogen peroxide activation and the new frontiers in this fascinating branch of the selenium chemistry, including Lewis base catalysis and selenium-π-acid catalysis. As the reader will note in the pages of this chapter, the combination of the Se-organocatalysis with the use of alternative green solvents (water, perfluorinated, ionic liquids, glycerol-based solvents, etc.), atom-economic reactions, use of green oxidants (H_2O_2, air), and solid supported catalysis makes Se-catalyzed reactions a feasible and robust green alternative in organic synthesis. The mechanism, scope, and limitations of the reactions displayed in the chapter are presented and discussed, giving subsidies to the reader for a clear understanding of the state of the art in the use of organoselenium compounds as a catalyst in organic synthesis.

1.1 Introducing the Organoselenium-Catalysis

Over the last decades, organoselenium has moved from an exotic, bad smelling chemistry, to an environmentally friendly tool to promote a greener organic synthesis. Despite it is not an easy journey and there is not a shortcut in the way to change this paradigm, some recent efforts of the selenium chemists' community have brought up remarkable behaviors of selenium-containing small molecules.

Since the discovery that selenium, in the form of selenocysteine (Sec), plays a central role in the activity of several enzymes, such as glutathione peroxidase (GPx), thioredoxin reductase (TrxR), and selenoprotein P [1–3], the search for low-molecular mass, biomimetic synthetic selenium-containing compounds has exponentially increased. Ebselen (2-phenyl-1,2-benzisoselenazol-3(2*H*)one) (Fig. 1.1) is probably the most famous member of the organoselenium family, whose notoriety is mainly due to its antioxidant activity [4–7]. The role of this and several other bioactive organoselenium compounds is discussed in Chap. 2 of this book. In this section, we show some examples of efficient bioinspired catalysis using organoselenium compounds as catalyst.

© Springer International Publishing AG, part of Springer Nature 2018

E. J. Lenardão et al., *New Frontiers in Organoselenium Compounds*,
https://doi.org/10.1007/978-3-319-92405-2_1

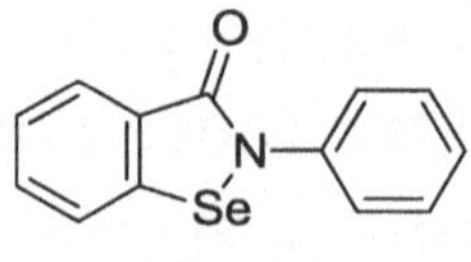

Fig. 1.1 Structure of ebselen

Reactions where the organoselenium compounds are used as activators or promoters, i.e., those where relatively large amounts (stoichiometric or over-stoichiometric ones) of the substance are required to trigger the reaction, will not be discussed in this chapter. In this book, an organoselenium catalyst is that compound used in sub-stoichiometric amounts. The use of selenides and diselenides as ligands in metal-catalyzed reactions will be not discussed here. For those interested in such reactions, several excellent book chapters and reviews are available [8–12].

The first examples of the use of organoselenium compounds as catalysts involved mainly oxygen-transfer reactions, by means of the activation of hydrogen peroxide [12–20]. As described in this chapter, however, over the last years several new reactions have been made possible thanks to the use of different approaches that put the organoselenium-catalyzed reactions in a prominent position, expanding their use to a variety of novel reactions.

The different facets of organoselenium catalyzed reactions can be divided according the type of reaction and/or the products obtained. Obviously, any attempt of classification of the Se-catalyzed reactions will bump into exceptions or intermediate situations, with characteristics of more than one category. Even if it risks being inaccurate, we will make a classification that we consider adequate for a better understanding of the role of the selenium catalyst in the reactions under discussion in this chapter.

The most common use of organoselenium catalysts is by far in oxygen-transfer reactions, followed by Se-π-acid catalysis (using electrophilic selenium species) and as Lewis base catalysts in the activation of electrophiles (Fig. 1.2). In the following sections we will present some representative examples of Se-catalyzed reactions to prepare functionalized compounds together the mechanism of the catalytic cycle, when it is available and relevant. The brand new chalcogen-bonding catalysis is discussed in Chap. 4, together with the discussion on such type of interaction.

1.2 Electrophilic Selenium as Alkene Activators by π-Acid Catalysis

The electrophilic selenium catalysis (ESC), or the selenylation–deselenylation of alkenes, is one of the most useful selenium-catalytic process, allowing the functionalization of non-activated alkenes [10, 11, 21–27]. In this reaction, the π bond of the alkene is activated by the selenium electrophile similarly to what occurs on activation with transition metals, with the π-orbital of the alkene interacting with the

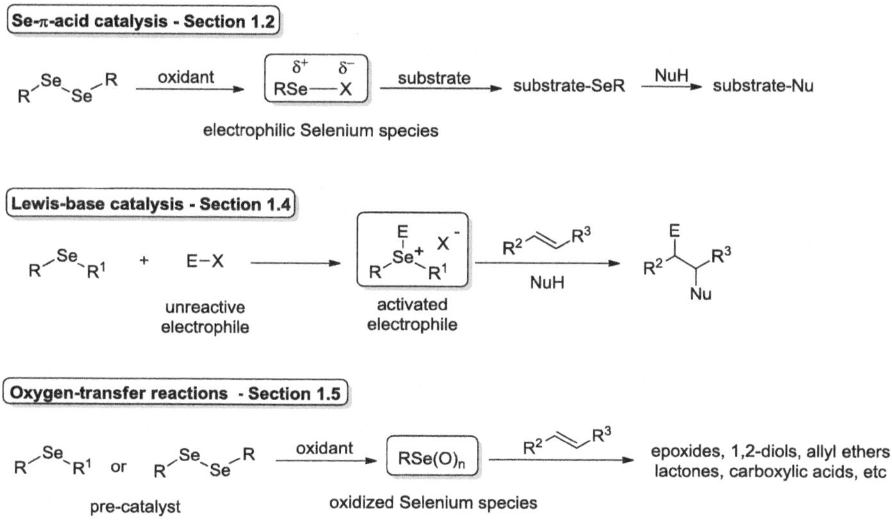

Fig. 1.2 Three main Se-catalyzed reactions

σ*-orbital of the selenium catalyst, combined with a back-donation to the olefinic π*-orbital by the chalcogen lone electrons [28]. The unstable, highly reactive seleniranium intermediate, can be trapped by nucleophiles, both endogenous (derived from the oxidant species) or exogenous.

Since the original work from Sharpless in 1979, which used a catalytic amount of compound **1a** to promote the allylic chlorination of olefins by *N*-chlorosuccinimide (NCS) [29, 30] (Scheme 1.1), a number of oxidants has been used in combination with different organoselenium species, including ammonium or sodium persulfate [31–38], hypervalent iodine [39–42] and hydrogen peroxide [43–48]. There are also few examples of electrochemically generated electrophilic selenium species as well [49–52].

The electrochemical selenylation–deselenylation reaction was firstly described by Torii and coworkers in 1980, using a stoichiometric amount of $(PhSe)_2$ **1a** in the presence of Et_4NBr (10 mol%) using Pt foils as electrodes [53]. In this reaction, molecular bromine was generated from Et_4NBr, allowing the formation in situ of the electrophile PhSeBr, which in turn reacted with a non-activated alkene. Once a mixture of H_2O/CH_3CN was used as the solvent, the respective allyl alcohol was obtained in 85% yield. The same group developed the first catalytic version of the electrochemical oxyselenylation–deselenyation of olefins to prepare allyl alcohols and ethers from isoprenoids [49]. In the pioneer work of Torii and coworkers, a mixture of the alkene, $(PhSe)_2$ **1a** (10 mol%) and $MgSO_4$ (2.5 equiv.) in H_2O/CH_3CN or MeOH as the solvent, was electrolyzed under a constant current (6.7 mA/cm^2; 3.1–11.9 F/mol) using Pt foils as electrodes at 66–68 °C. The expected allyl alcohols or allyl ethers were obtained in 63–92% yields.

The Torii's discovery was revisited 25 years later by Wirth and coworkers, who developed a catalytic version of the aforementioned original stoichiometric reaction

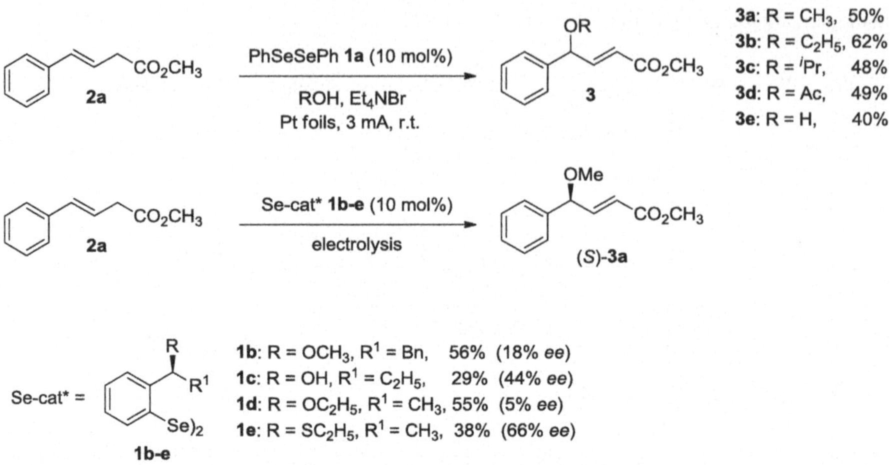

Scheme 1.1 Catalytic selenylation-deselenylation reactions

using **1a**/Et_4NBr [51]. The use of a small amount of H_2SO_4 and a higher current (3 mA) were mandatory to allow the use of catalytic amounts of **1a** in the electrochemical reaction. The authors were able to prepare different allyl ethers or alcohols, derived from (*E*)-methyl 4-phenylbut-3-enoate **2a** in moderate yields (40–62%) just by changing the solvent (methanol, ethanol, *iso*-propanol, or water). When acetic acid was used as the solvent, $(NH_4)_2S_2O_8$ was used instead Et_4NBr, affording the acylated product **3d** in 49% yield (Scheme 1.1). The method was successfully extended to (*E*)-4-phenylbut-3-enenitrile instead ester **2a**, but did not work when β-methyl styrene was used as the alkene. Four different chiral diselenides **1b–e** were tested in the asymmetric allylic methoxylation of **2a**, affording preferentially (*S*)-**3a** (29–56% yields; 5–66% *ee*). The best yield (56%) was obtained using diselenide **1b**, while the highest *ee* (66%) was delivered by catalyst **1e** (5 mol%), containing a sulfur atom.

A plausible mechanism of this reaction involves firstly the anodic oxidation of Br^- to molecular bromine, which reacts with $(PhSe)_2$ **1a** to form PhSeBr, the electrophile of the reaction. The reaction of PhSeBr with alkene **2a** forms the seleniranium cation **A** that undergoes a nucleophilic attack from the solvent, affording selenide **B**. Then, the electrochemically generated bromine reacts with **B**, forming the unstable tetravalent selenylbromide **C**, that delivers the desired allylic ether **3a** after the elimination of HBr and PhSeBr ready to start a new catalytic cycle (Scheme 1.2).

The use of persulfate ($S_2O_8^{2-}$) as an oxidant to generate the selenium electrophile in ESC reactions was firstly explored by Tiecco and coworkers [37], and independently, by Tomoda and Iwaoka [35]. These works started to pave the road where ESC is currently being developed, with new oxidants and cooperative catalysis, opening new frontiers in this exciting journey. Most of the works from Tiecco and Tomoda on the synthesis of allyl ethers, alcohols and esters have been previously discussed in several book chapters [20–22] and reviews [13–19] and will be not covered here.

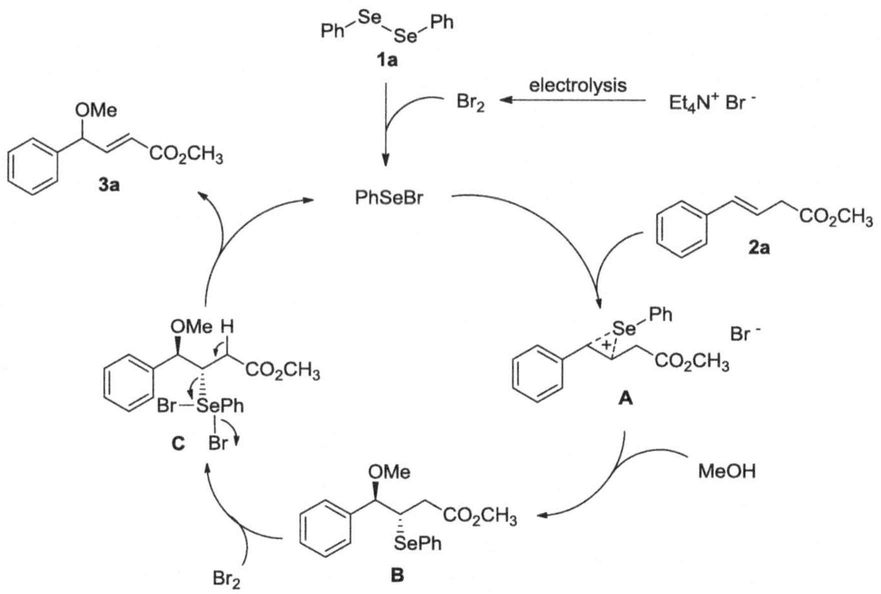

Scheme 1.2 Proposed mechanism for the catalytic selenylation-deselenylation

Successful asymmetric versions of this selenylation–deselenylation reaction using chiral diselenides **1f–n** were developed by Tomoda (**1f**/$Na_2S_2O_8$) [54], Fukuzawa (**1g**/$(NH_4)_2S_2O_8$) [55], Wirth (**1h–l**/$Na_2S_2O_8$ or $K_2S_2O_8$) [56] and Tiecco (**1m–n**/$(NH_4)_2S_2O_8$) [57, 58] (Fig. 1.3). Catalysts **1f** and **1h–l** were used in the asymmetric methoxylation of β-methylstyrenes, affording the respective (*R*)-allyl ether as the major enantiomer after 7 days of reaction. The reaction of β,γ-unsaturated esters in the presence of **1g** or **1m–n**/$(NH_4)_2S_2O_8$ and ethanol or methanol as the solvent, afforded the respective γ-alkoxy-α-β-unsaturated esters after 7 days (for **1g**) or 2–4 days (for **1m–n**) of reaction at room temperature with *ee* from 4% (**1g**) to 82% (**1n**).

A remarkable advance in the ESC occurred after the discovery that some *N*-fluorinated heterocycles were able to work both as final oxidant and endogenous nitrogen source in allylic and vinylic amination reactions [23, 24]. Inspired by the seminal works of Sharpless [29, 30] and Tunge [59], Breder and coworkers designed a catalytic system using *N*-fluorobenzenesulfonimide (NFSI) as the oxidant and diphenyl diselenide **1a** as the catalyst [60]. By changing NCS for NFSI, the authors could prepare allylic imides **5** starting from β,γ-unsaturated esters, sulfones, amides, phosphates, nitriles, and ketones **4**. When cyclic alkenes **6** were used, however, vinyl amides **7** were preferentially formed, showing for the first time the possibility of using ESC in oxidative C_{sp2}-H imidation. Cyclohexenone, sulfolene, cyclopentene, and styrene derivatives were successfully used as substrates to prepare the respective vinyl imides (Scheme 1.3).

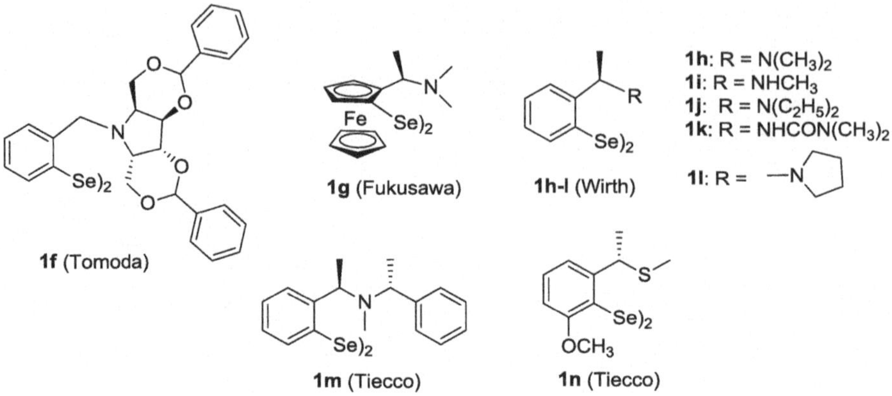

Fig. 1.3 Chiral diselenides used as catalysts in selenylation-deselenylation reactions

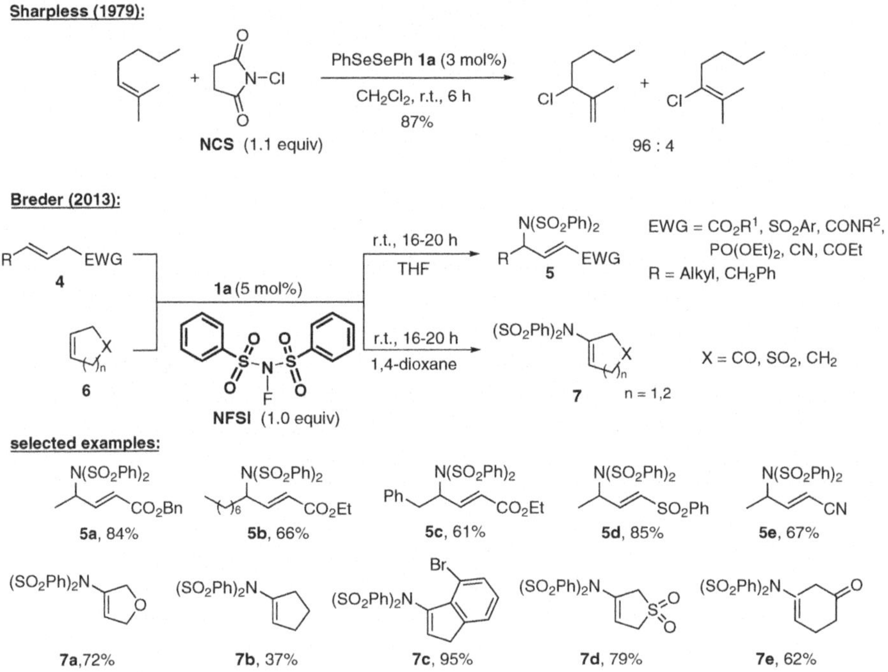

Scheme 1.3 Allylic and vinylic chlorination and amination

The authors proposed a catalytic cycle for the allylic and vinylic amination catalyzed by $(PhSe)_2$ (Scheme 1.4), in which the first step is the nucleophilic attack of diselenide **1a** to NFSI to form the highly electrophilic cationic species **A** and fluoride anion. Following, **A** reacts with the alkene **4** or **6** to form the cationic adduct **B**, which undergoes elimination assisted by F^-, regenerating diphenyl diselenide **1a** and forming the allyl **5** or vinyl imides **7** [60].

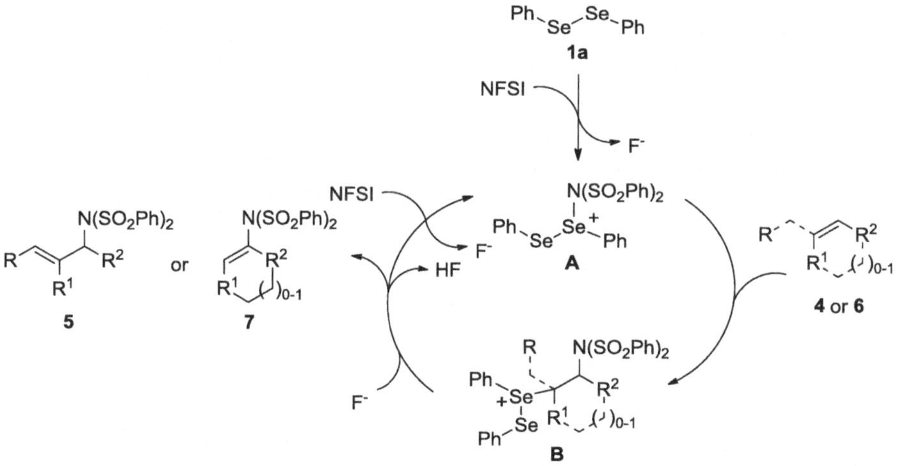

Scheme 1.4 Catalytic cycle for the allylic and vinylic amination catalyzed by $(PhSe)_2$

Two years later, Zhao and coworkers reported a ESC, hydroxy-controlled amination of terminal allylic alcohols **8** using NFSI as the final oxidant [61]. The authors observed that a base is mandatory to stop the reaction in the vinylic amination stage, once in the presence of acid the hydrolysis to the respective α,β-unsaturated aldehydes occurs. This protocol can be tuned to selectively prepare enamines **9** and aldehydes **10** by using or not a base (NaF) in the reaction media (Scheme 1.4). The method was successfully used in the oxidative amination of α,β-disubstituted allyl alcohols **11**, giving the respective *cis*-amino derivatives **12** in 71–92% yields after 12 h of reaction in THF. In this reaction, no base was necessary, once amines **12** are more stable compared to unsubstituted analogues **9**, and no aldehyde formation was observed (Scheme 1.5).

A mechanism for the reaction was proposed, which involves firstly the reaction of $(PhSe)_2$ **1a** with NFSI to produce the highly electrophilic selenium species **A** (The authors did not discard the possible formation of $PhSeN(SO_2Ph)_2$ and PhSeF (Scheme 1.6)). In the presence of allylic alcohol **8h** (R = Ph, R^1 = H), **A** could interact with its hydroxyl group, giving the aziridinium **B** or selenonium adduct **C**. The lone pair of electrons from oxygen of hydroxyl in **8h** is important to stabilize these onium ions. The catalyst $(PhSe)_2$ is then quickly eliminated to give intermediate **D**, which in the presence of base delivers the 3-amino allylic alcohol **9h**. In the absence of base, **9h** is converted to the unsaturated aldehyde **10h**. The high selectivity of the amination is attributed to the assistance of hydroxyl group and the fast formation of **D** that inhibit the formation of 2-aminated product and 3-amino 1-phenyl propanone [61].

In the same year, two quite similar protocols to prepare indole derivatives **15** and **16** were developed by Breder [62] and Zhao [63] almost simultaneously, involving the reaction of *o*-vinyl anilines **13** in the presence of NFSI and a catalytic amount of $(PhSe)_2$ **1a** (Scheme 1.7). Both protocols differ slightly in the amount of catalyst and

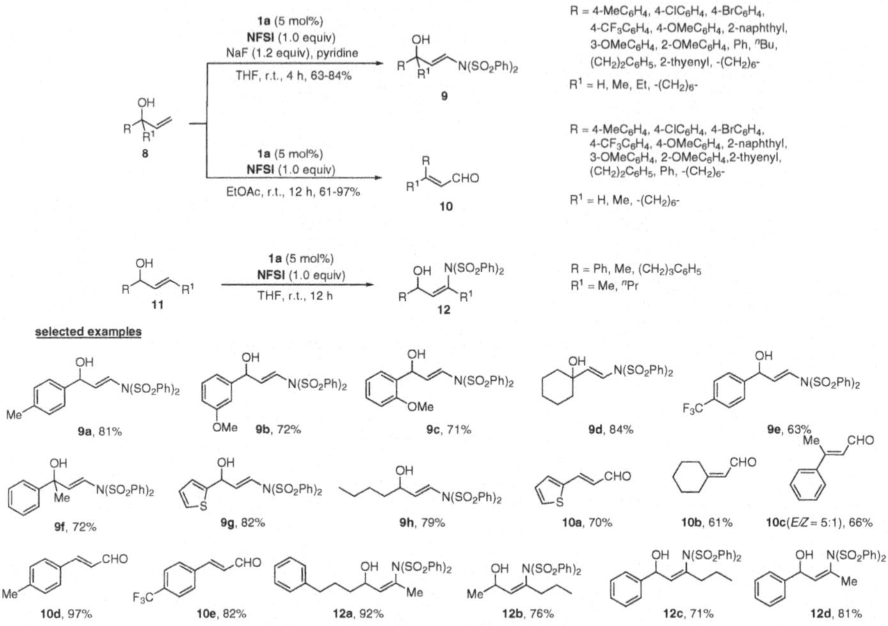

Scheme 1.5 Amination of terminal and disubstituted allylic alcohols using NFSI as final oxidant

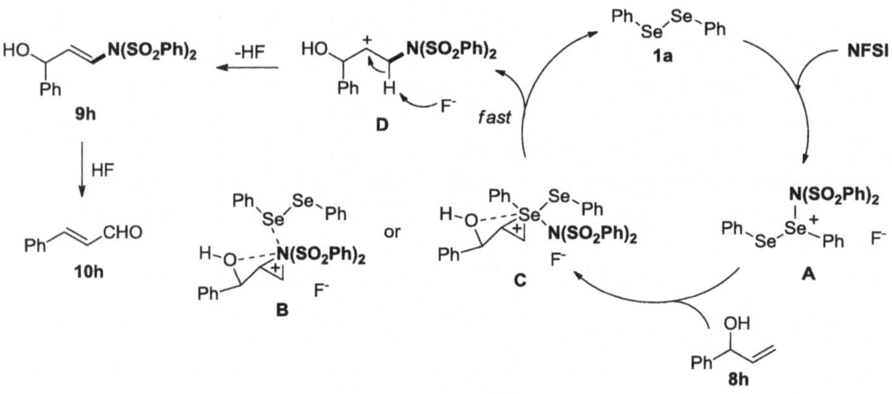

Scheme 1.6 Proposed mechanism for the NFSI-promoted amination

the solvent: 2.5 mol% of **1a** in toluene/4 Å MS at 100 °C for 16 h [62] or 10 mol% of **1a** in 1,4-dioxane at 30 °C for 18 h [63]. The Breder's protocol was successfully extended to aminopyridines **14**, but a larger amount of catalyst **1a** (5 mol%) was necessary in comparison to the anilines **13**.

A catalytic cycle was proposed by Breder, which firstly involves the oxidative cleavage of the Se-Se bond of diphenyl diselenide **1a** by NFSI to form the electrophilic species of selenium, which reacts with the alkene **13a** to give the seleniranium

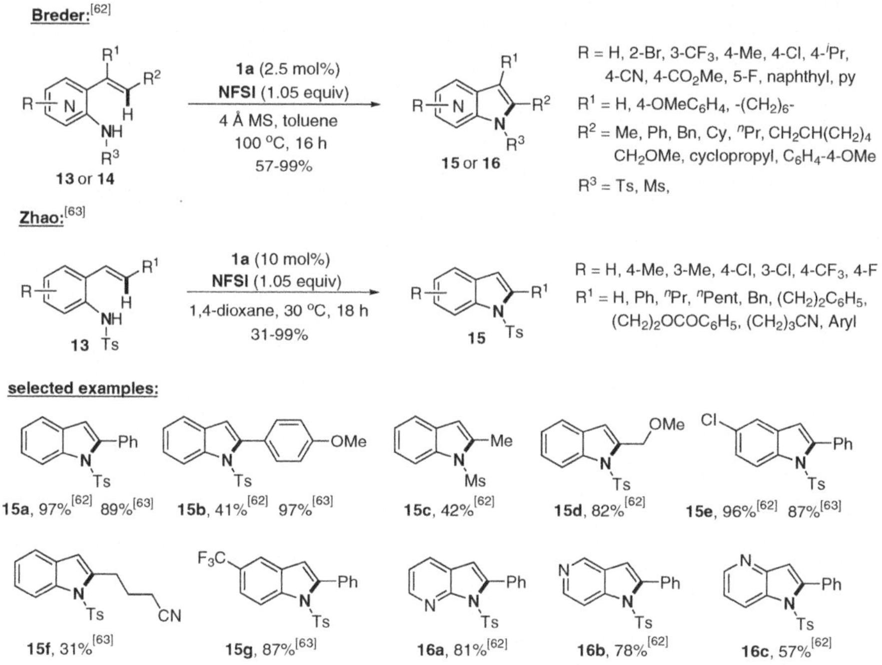

Scheme 1.7 Synthesis of indoles developed by Breder and Zhao

ion **A** [62]. Following, an intramolecular cyclization occurs, through an attack by the protected adjacent *N*-atom in **A**, affording the selanyl indoline **B**. The reaction with a second electrophilic selenium species activates the PhSe group for an elimination step in intermediate **C**, releasing the desired indole **15a** and $(PhSe)_2$ **1a** for a new catalytic cycle (Scheme 1.8).

An intramolecular version of ESC *O*- and *N*-allylation was described by Zhao and coworkers using 1-fluoropyridinium triflate (FP-OTf) as the final oxidant and $(PhSe)_2$ as catalyst (Scheme 1.9) [64]. In the optimization studies using unsaturated alcohol **17a** as starting material, 10 mol% of $(PhSe)_2$ **1a** as a catalyst and NFSI (1.2 equiv.) as oxidant in acetonitrile, the desired tetrahydrofurane **18a** was obtained in 59% yield, together the rearrangement product tetrahydrooxepine **24a** (19%) after 18 h at room temperature. By changing the oxidant by FP-OTf and adding NaF (1.0 equiv.) as a base, the yield of **18a** increased to 98% and no product of rearrangement was observed. The optimized condition was extended to several properly substituted unsaturated alcohols **17** or sulfonamides **20** to deliver tetrahydrofurans **18**, tetrahydropyrans **19**, pyrrolidines **21**, piperidines **22** (Scheme 1.9). The better solvent for the *N*-cyclization reactions was DCE and the azepines **23** were selectively prepared using 0.5 equiv. of NaF. The authors rationalized the formation of **23** by a HF-promoted rearrangement of the parent pyrrolidines **21**.

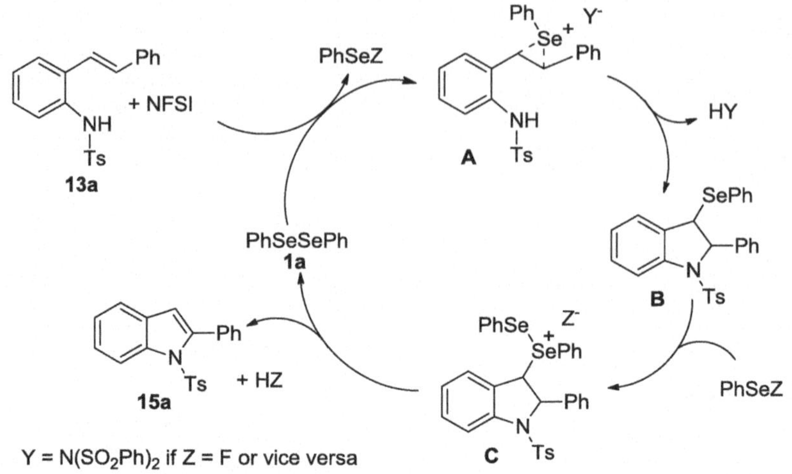

Scheme 1.8 Catalytic cycle proposed by Breder to prepare indoles from *o*-vinyl anilines

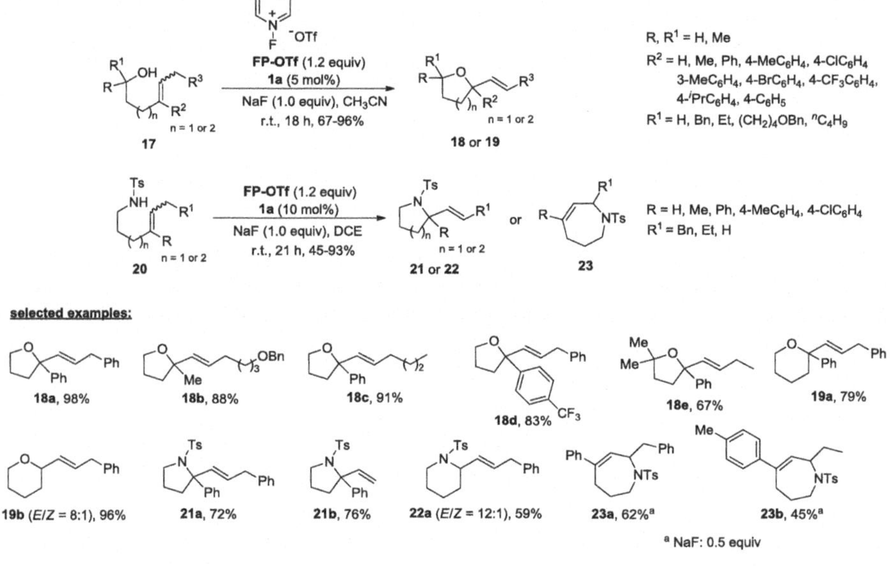

Scheme 1.9 Intramolecular ESC of *O*- and *N*-allylation described by Zhao

The authors conducted a series of designed control reactions to verify the role of FP-OTf in the *O*-cyclization reaction of **17a**. They concluded that first, the reaction proceeds through a Se-addition and elimination sequence and second, that FP-OTf is essential in the elimination step, by activating the PhSe group of the intermediate **B**. A mechanism for formation of **18a** from **17a** catalyzed by $(PhSe)_2$ **1a** is showed in Scheme 1.10, which is quite similar to that of the indole formation

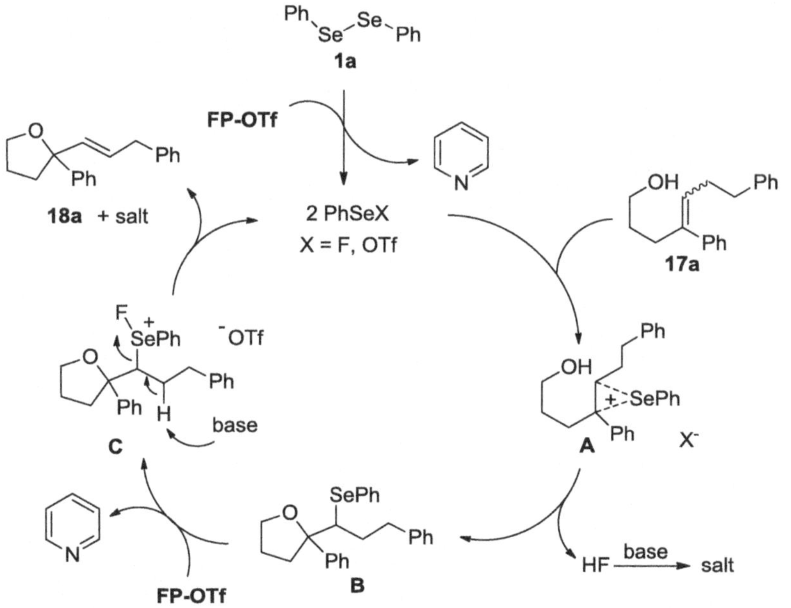

Scheme 1.10 The role of 1-fluoropyridinium triflate (FP-OTf) in the *O*-cyclization reaction

in the presence of NFSI, depicted in Scheme 1.8. The first step is the formation of the highly reactive selenium species PhSeX (X = F or OTf), which adds to alkene double bond of **15a**, forming the seleniranium **A**. Then, the three-membered ring is opened by the hydroxyl group, forming the intermediate tetrahydropyran **B**. Following, FP-OTf oxidizes the SePh group, giving intermediate **C**, delivering the desired product **18a** by releasing of PhSeX to a new catalytic cycle. This step is facilitated by the presence of the base.

An efficient and selective Se-catalyzed direct C-2 C-H pyridination of 1,3-dienes **24** was developed by Zhao and coworkers, making this important transformation feasible in a general way for the first time [65]. The authors started their studies using $(PhSe)_2$/FP-OTf as a catalyst/oxidant system, however unsatisfactory outcomes were obtained. After a detailed screening of different diselenides and oxidants, dibenzyl diselenide **1o**, combined with FP-OTf, showed the best results (Scheme 1.11).

During the investigation on the reaction mechanism, the authors observed the formation of *N*-benzylpyridinium salt **E** as a decomposition product of dibenzyl diselenide **1o**, together the selenylated intermediate **C**, indicating that the reaction involves a selenylation-deselenylation pathway. The seleniranium intermediate **A** is stabilized by the conjugated double bond, contributing to structure **B**. The selective attack of pyridine to **A** gives **C**, which is oxidized by FP-OTf, forming the salt **D**. The elimination of HF releases the desired product **25a** and BnSeX for a new catalytic cycle (Scheme 1.12).

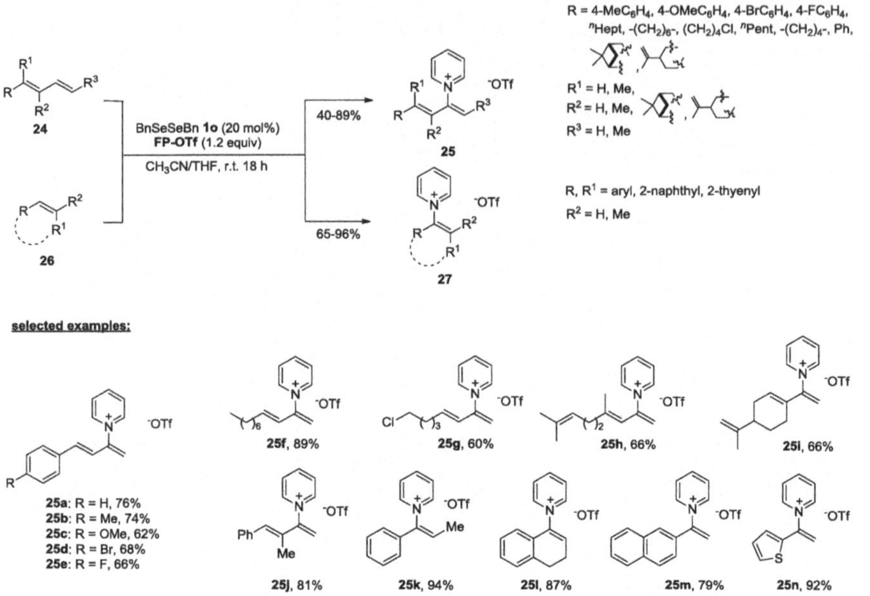

Scheme 1.11 Se-catalyzed C-2 C-H pyridination of 1,3-dienes using FP-OTf as final oxidant

By changing FP-OTf for *N*-fluoro-2,4,6-trimethylpyridinium tetrafluoroborate, [TMPyF][BF_4], and using Selectfluor® as a co-oxidant, different pyridines **28a–d** were efficiently used as a nitrogen source, affording the respective pyridinium salts **29a–d** in 53–86% yields (Scheme 1.13). An intramolecular version of this reaction was achieved starting from pyridinyl 1,3-dienes **30a–b**. In this case, new six-membered-fused pyridinium salts **31a–b** were prepared in good yields [65].

Breder developed a Se-catalyzed C_{sp3}-H bond acyloxylation to prepare isobenzofuranones **33** by a 5-*exo*-trig cyclization of *o*-allyl benzoic acids **32** [66]. Diphenyl diselenide **1a** (10 mol%) was used as the catalyst and NFSI was the final oxidant (Scheme 1.14). It was observed a negative influence of electron-withdrawing CF_3 group in the aromatic ring conjugated to the double bond, as in **33d**, and only 6% of the desired product was obtained. Interestingly, the selectivity was switched to 6-*exo*-trig cyclization in **32g**, with a *n*-propyl group instead an aromatic one, affording the isochroman-1-one **33g**, instead the expected isobenzofuranone.

The mechanism of the reaction was investigated, and among three possible scenarios, the authors proposed that the seleniranium **A** is firstly formed by the reaction of the electrophilic selenium species [Se] with *o*-allylic benzoic acid **32a**. Then, a proton abstraction occurs, with opening of the three-membered ring to form selenonium ion **B**, which suffers an intramolecular attack by the carbonyl oxygen, delivering the desired isobenzofuranone **33a** and the selenium species for a new reaction (Scheme 1.15).

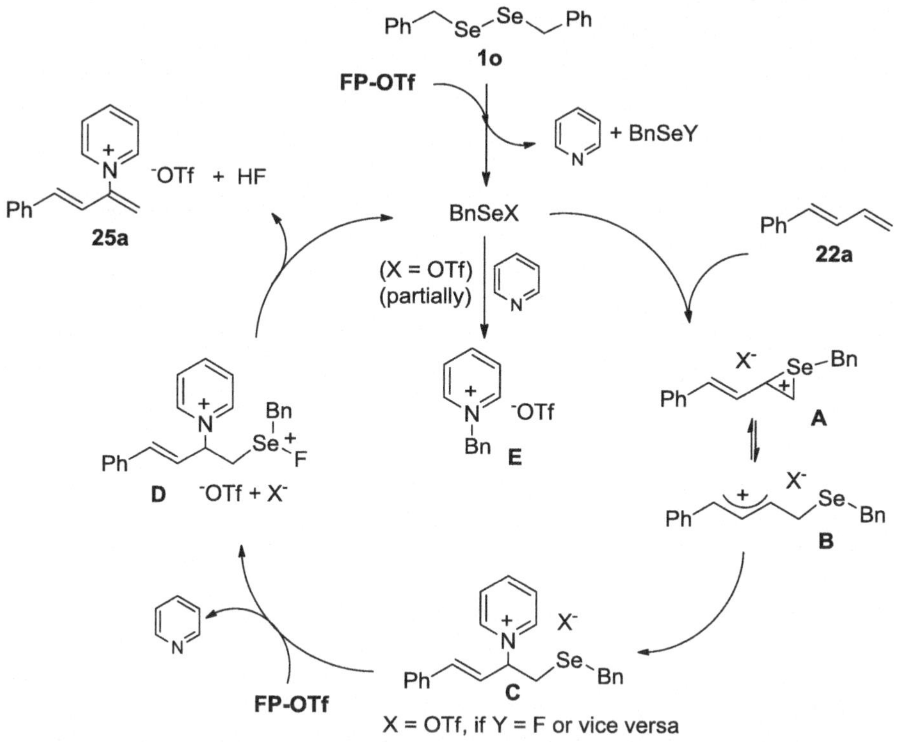

Scheme 1.12 Proposed mechanism for the Se-catalyzed C-2 C-H pyridination

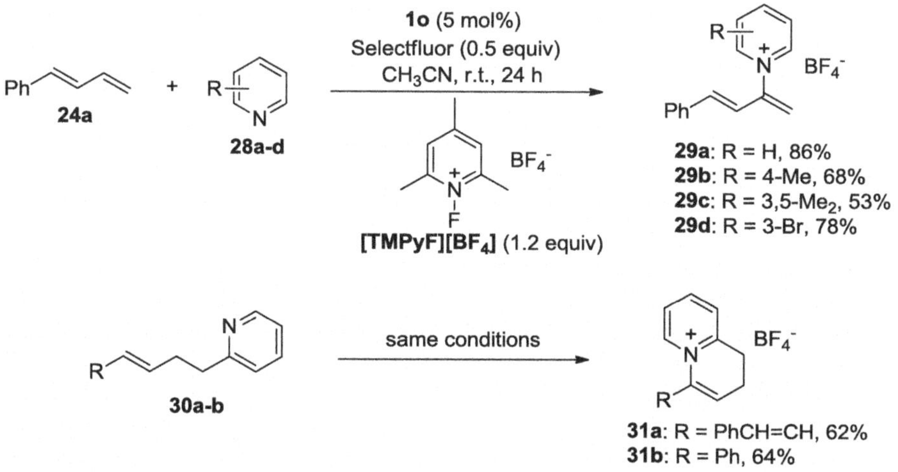

Scheme 1.13 Se-catalyzed C-2 C-H pyridination of 1,3-dienes using [TMPyF][BF_4] as final oxidant

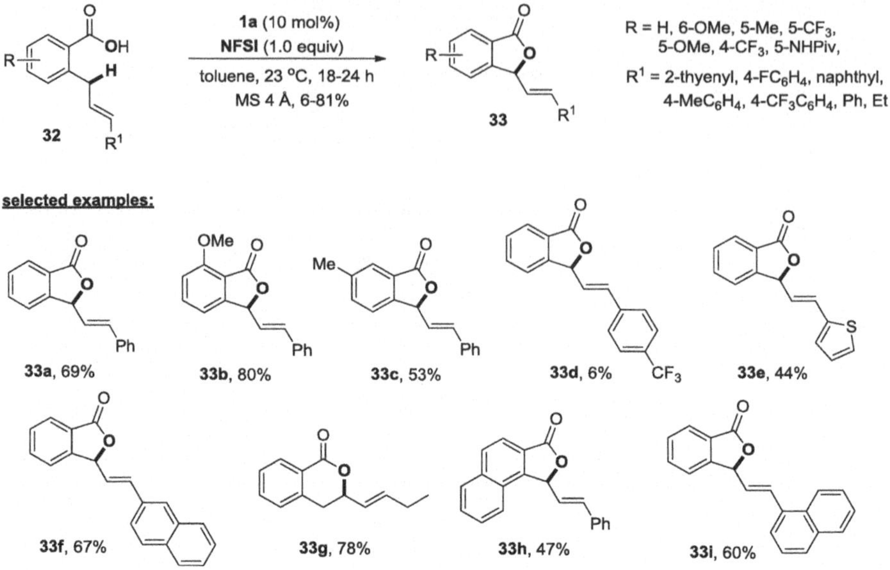

Scheme 1.14 Se-catalyzed C_{sp3}-H bond acyloxylation to prepare isobenzofuranones

Maruoka and coworkers described an asymmetric version of the Se-catalyzed C_{sp2}-H acyloxylation starting from β,γ-unsaturated acids **34** using a new indanol-derivative chiral selenide **35**, which was activated by NFSI as the oxidant [67]. By this procedure, several enantioenriched γ-butenolides **36** were smoothly obtained with up to 97% *ee* and from 60% to 99% yields (Scheme 1.16). The initial idea was to prepare the respective diselenide derived from 6-metoxyindanone, however experimental drawbacks in the purification step led the authors to designing a chiral selenide that could generate the electrophilic selenium catalyst in situ in the presence of an oxidant. The authors observed that the absence of $CaCO_3$ or reducing the temperature to 0 °C, caused a decrease in both yield and enantioselectivity of the reaction. The method showed to be robust, being suitable to a diversity of functionalized γ-alkyl- and γ-aryl-substituted alkenoic acids **34**. In the case of the aryl derivatives, $TMSOCOCF_3$ was used instead $CaCO_3$ and the temperature was reduced to 10 °C for improving the enantioselectivity (Scheme 1.16).

Among the different tested oxidants ($PhI(OCOCF_3)_2$, $Na_2S_2O_8$), NFSI was the only capable of oxidize the selenide **35** to the electrophilic species **35***, giving excellent yields after only 2 h of reaction (Scheme 1.17).

A cooperative interaction between diphenyl diselenide **1a** and the photoredox catalyst 2,4,6-tri(4-methoxyphenyl)pyrylium tetrafluoroborate, (*p*-anisyl)pyr **39**, allowed the use of air or molecular oxygen as terminal oxidants in the oxidative esterification of alkenes [68]. By this protocol, functionalized and non-functionalized alkenes **37** were reacted with carboxylic acids **38** and converted to the respective allylic esters **40** in up 89% yield, in excellent regioselectivity and good functional group tolerance (Scheme 1.18).

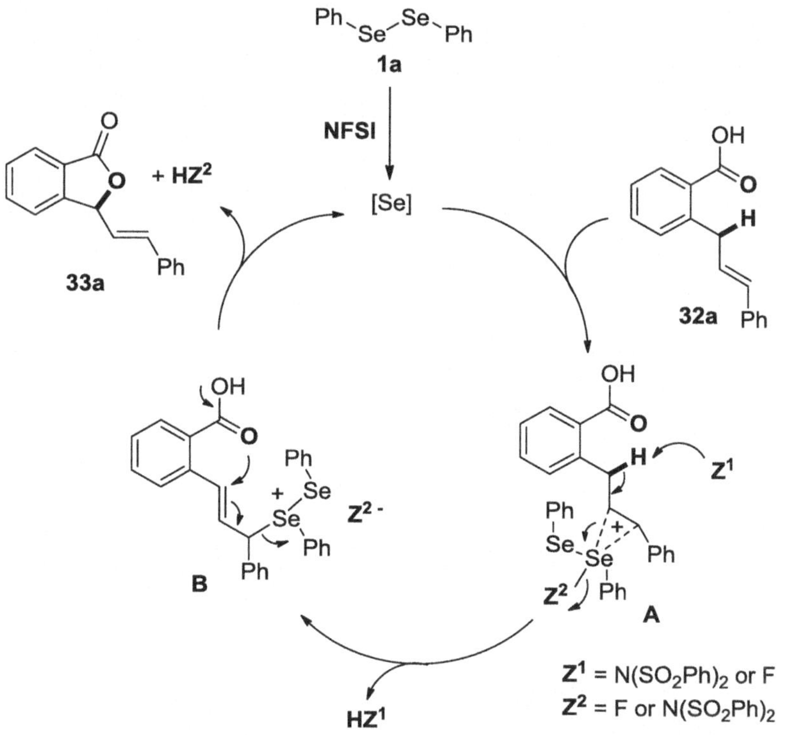

Scheme 1.15 Proposed mechanism for the Se-catalyzed cyclization of *o*-allylic benzoic acid

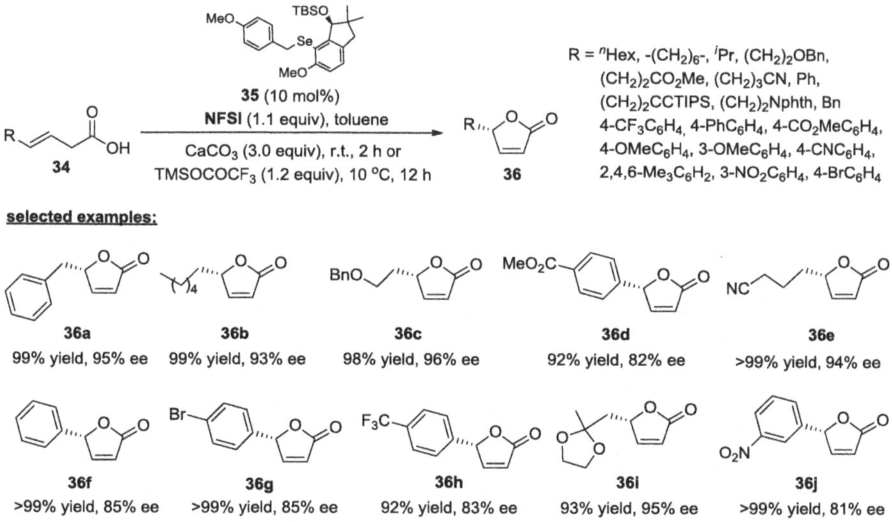

Scheme 1.16 Asymmetric Se-catalyzed C_{sp2}-H acyloxylation of β,γ-unsaturated acids

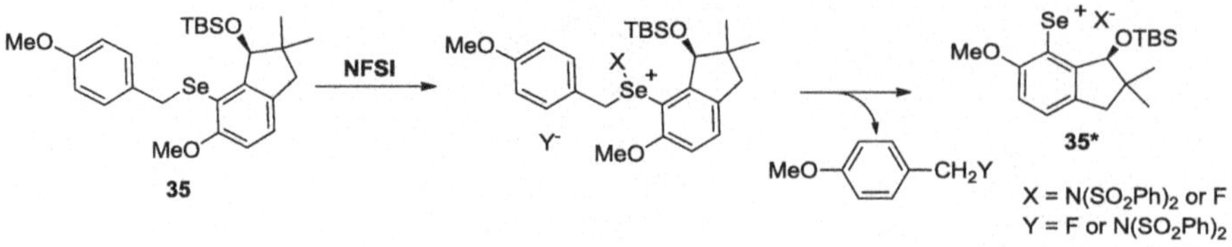

Scheme 1.17 Formation of the electrophilic species **35*** from selenide **35** and NFSI

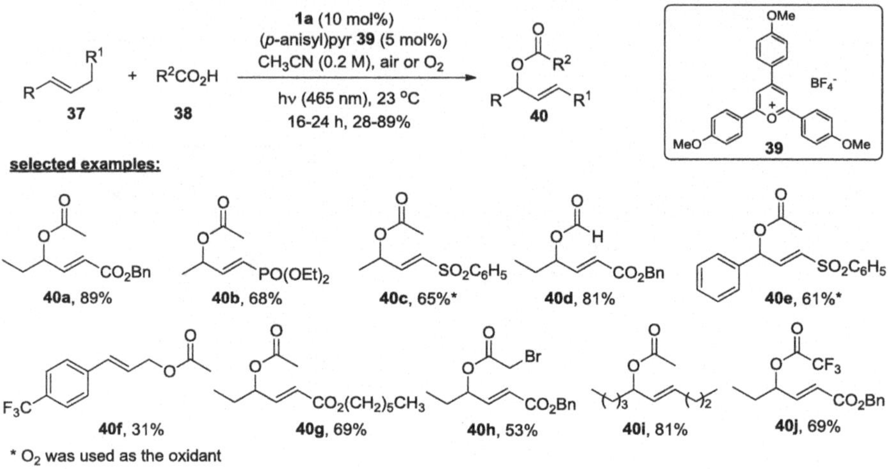

Scheme 1.18 Air or molecular oxygen as terminal oxidants in the Se-catalyzed esterification of alkenes

A mechanism for this cooperative Se-catalyzed oxidative esterification of alkenes was proposed after a series of control experiments and it is depicted in Scheme 1.19 for the synthesis of **40a**. The photoredox catalytic cycle is responsible by the formation of the highly electrophilic selenium species **B**, via single electron transfer (SET) from **1a** to the excited photocatalyst **39***. Intermediate **B** reacts with alkene **37a** to form the seleniranium **C** and HA. Following, acetic acid **38a** acts as a nucleophile, opening the seleniranium ring to give intermediate **D** which, in the presence of the excited photocatalyst **39*** delivers the radical cation **E**, by SET. Finally, the PhSe group is eliminated to regenerate **1a** and give the desired ester **40a** in the Se-catalytic cycle. The reduced photocatalyst **A** is oxidized by molecular oxygen to regenerate **39** for a new photocatalytic cycle.

The Se-catalyzed oxidative cyclization was used as strategy in the synthesis of the (+)-Greek tobacco lactone **43**, a natural-occurring bicyclic C_{11}-homoterpenoid, using $(PhSe)_2$ **1a** as the catalyst and either NFSI or molecular oxygen as the terminal oxidant [69]. The key intermediate **42** was prepared from (*R*)-linalool **41** in three steps (Scheme 1.20). By using the **1a**/NFSI system in THF and in the presence of $CaCO_3$ and 4 Å molecular sieves, lactone **43** was obtained as the sole diastereoisomer in 60% yield after 3 h at room temperature. Alternatively, lactone **43** was prepared

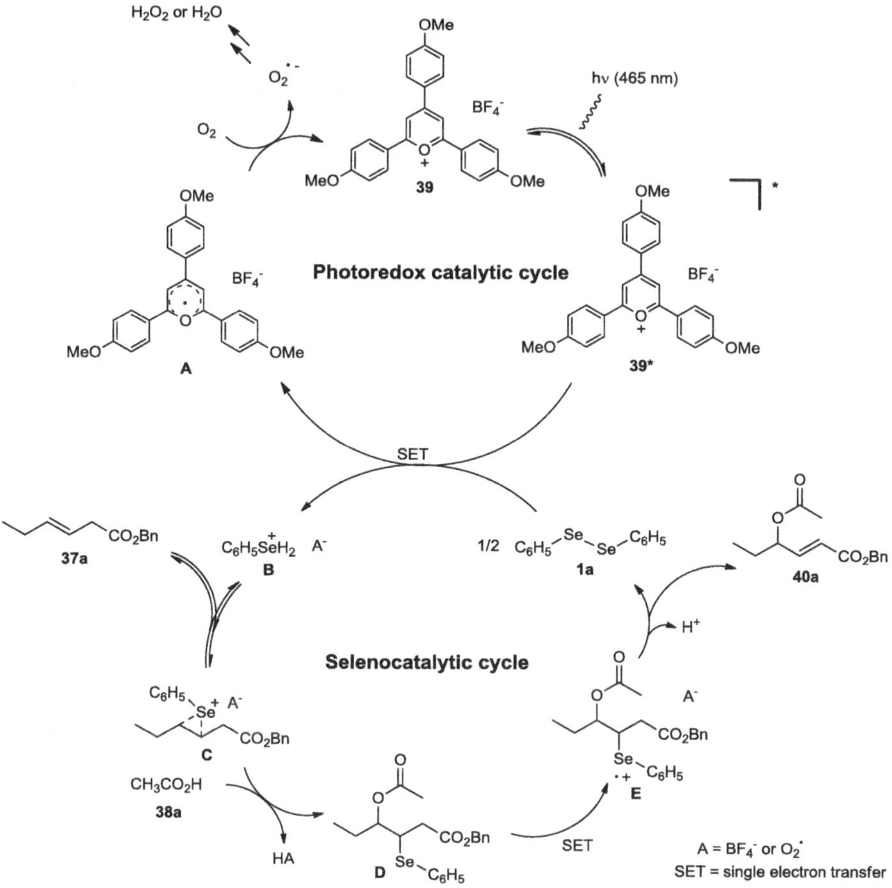

Scheme 1.19 Proposed mechanisms for the photoredox and the selenocatalytic cycles

in 83% yield (dr = 84:16) using the same strategy of Scheme 1.18, i.e., diphenyl diselenide **1a** (10 mol%) combined with photocatalyst **39** (5 mol%) in the presence of light (465 nm) and molecular oxygen [68]. In an optimized condition, Na_2HPO_4 (0.8 equiv.) was added and the solvent was changed from CH_3CN to DCE. The same cyclization reaction of **42** was proved using Pd catalysis; however, a mixture of the desired lactone **43** (41% yield, dr = 85:15) and Δ^6-tetrahydrooxepine **44** (15% yield) was obtained, highlighting that the Se-catalyzed protocol showed to be an efficient alternative to the Pd-catalyzed alkoxylation of nonactivated alkene **42**.

Besides electrophilic *N*-F reagents, such as NFSI, [TMPyF][BF_4], [TMPyF][BF_4], and Selectfluor®, stoichiometric amounts of hypervalent iodine have been used as an oxidant to generate the electrophilic selenium species in the Se-catalyzed intramolecular cyclization of alkenoic acids [40–42, 70]. Wirth and coworkers used bis(trifluoroacetoxy)iodobenzene (PIFA) to oxidize $(PhSe)_2$ **1a** to an electrophilic, highly reactive selenium reagent, that efficiently catalyzed the intramolecular

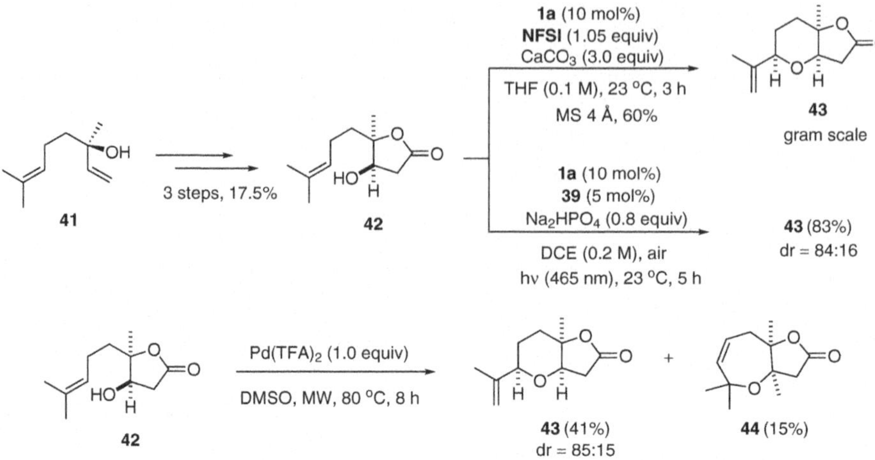

Scheme 1.20 Synthesis of the (+)-Greek tobacco lactone **43**, using $(PhSe)_2$ **1a** as the catalyst

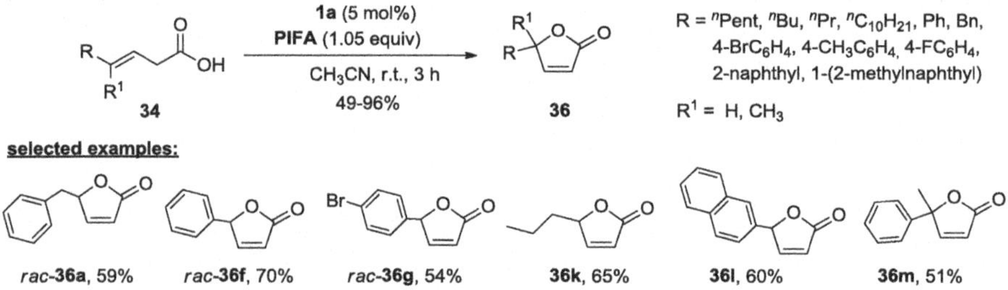

Scheme 1.21 Se-catalyzed intramolecular cyclization of β,γ-unsaturated carboxylic acids using PIFA as an oxidant

cyclization of β,γ-unsaturated carboxylic acids **34** to deliver the corresponding butenolides **36** in good yields after 3 h (Scheme 1.21) [42, 70]. Six different hypervalent iodine reagents and solvents were tested as the oxidant and the best yields were obtained using PIFA in CH_3CN in the presence of **1a**.

An asymmetrical version of the reaction using enantiomerically pure diselenides was unsuccessful, with longer reaction times, lower yields, and poor enantioselectivity [42]. To better results, a stoichiometric amount of diselenide and temperatures as low as −100 °C were necessary [71].

The proposed catalytic cycle for the reaction involves the preliminary formation of phenylselenenyl trifluoroacetate **A**, by the reaction of **1a** with PIFA. Then, reaction of **A** with alkenoic acid **34** gives the selenolactone **B**, which reacts with PIFA to afford the selenonium salt **C**. Following, butenolide **36** is formed by an elimination, while regenerating the selenium electrophile **A** (Scheme 1.22).

When γ,δ-unsaturated carboxylic acids **45** were used instead the β,γ-unsaturated ones, the respective 3,6-dihydro-2*H*-pyran-2-ones **46** were obtained in reasonable to good yields [40]. Among the different oxidants and solvents that

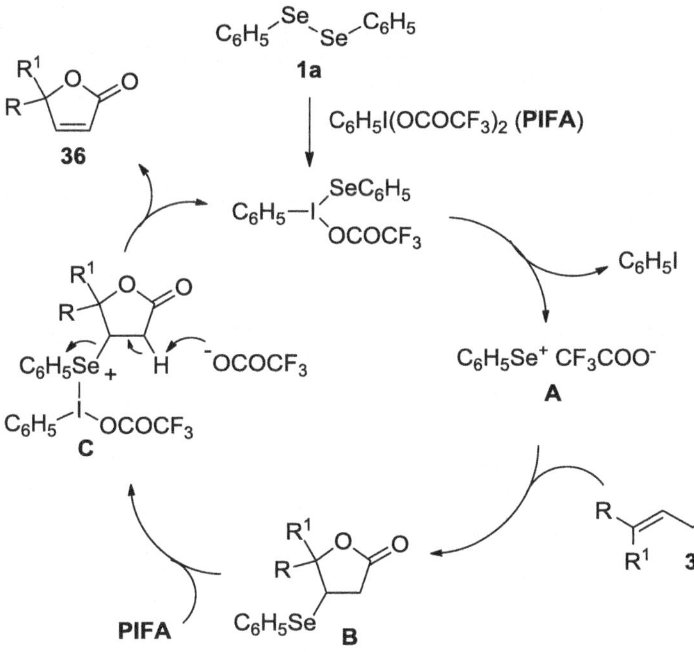

Scheme 1.22 The role of bis(trifluoroacetoxy)iodobenzene (PIFA) in the Se-catalyzed reaction

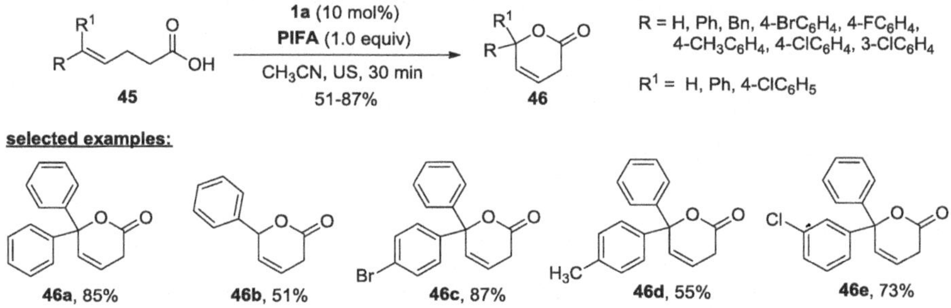

Scheme 1.23 Se-catalyzed intramolecular cyclization of γ,δ-unsaturated carboxylic acids using PIFA as an oxidant under ultrasound

were tested in the reaction, PIFA and CH_3CN showed the best results. Additionally, the authors observed that the irradiation with ultrasound (US) for 30 min slightly increased the reaction yields in the presence of $(PhSe)_2$ **1a** (10 mol%) as the catalyst (Scheme 1.23).

In the case of hexenoic acids **47a–d**, best results were obtained by stirring the reaction mixture at room temperature for 3 h instead using US. Four differently substituted γ-valerolactones **48a–d** containing both electron-withdrawing or electron-releasing groups at the aromatic ring were prepared by this protocol and better result was obtained when electron-releasing group ethyl was present in the

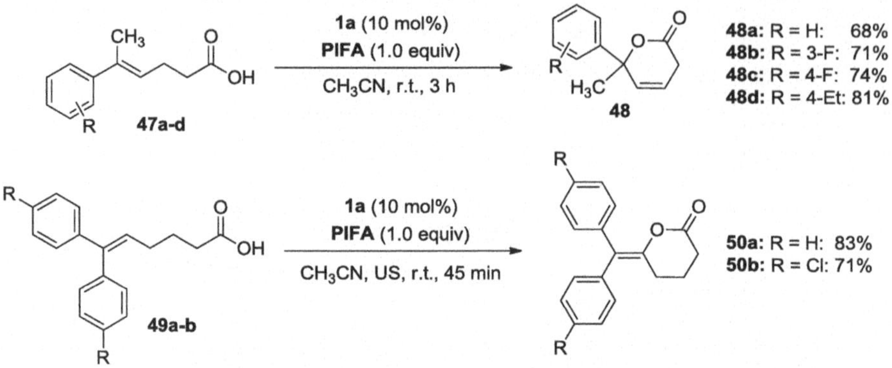

Scheme 1.24 Se-catalyzed cyclization of unsaturated carboxylic acids at r.t. or using US

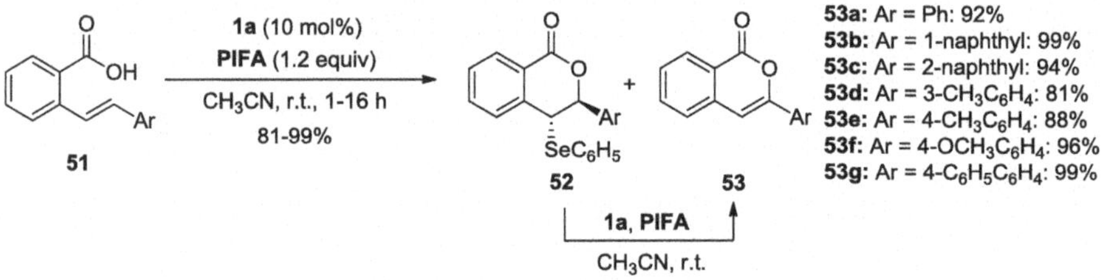

Scheme 1.25 Se-catalyzed synthesis of isocoumarins using PIFA as oxidant

benzene ring (Scheme 1.24). The method was extended to the cyclization of δ,ε-unsaturated carboxylic acids **49a–b** under US for 45 min. Contrary to expected, seven-membered ε-caprolactones were not observed but the six-membered 6-(diarylmethylene)tetrahydro-2*H*-pyran-2-ones **50a–b**, formed by an *exo*-cyclization, were isolated in 80% and 71% yields, respectively (Scheme 1.24) [40]. The mechanism proposed by the authors is quite similar to that for the cyclization of β,γ-unsaturated acids, showed in Scheme 1.22.

The same protocol was efficiently applied in the Se-catalyzed synthesis of isocoumarins **53**, through the 6-*endo-trig* cyclization of stilbenecarboxylic acids **51** in the presence of $(PhSe)_2$ **1a** and PIFA [74]. In some cases, the selenium-containing dihydroisocoumarin derivatives **52** could be isolated as minor side products. For most of the reactions, the initially formed dihydroisocoumarin **52** can be totally converted to the respective isocoumarin **53** just by extending the reaction time to up 16 h at room temperature (Scheme 1.25). The mechanism proposed for this reaction was the same of Scheme 1.22, for the synthesis of butenolides **36** from β,γ-unsaturated acids **34**, which involves the reaction of diselenide **1a** with PIFA to form the reactive electrophilic selenenyl trifluoroacetate.

In 2015, Denmark and coworkers described the first Se-catalyzed *syn*-stereospecific dichlorination of alkenes [72]. The authors used $(PhSe)_2$ **1a** (5 mol%) as a pre-catalyst, $BnEt_3NCl$ as chloride source and FP-BF_4 as the oxidant to prepare

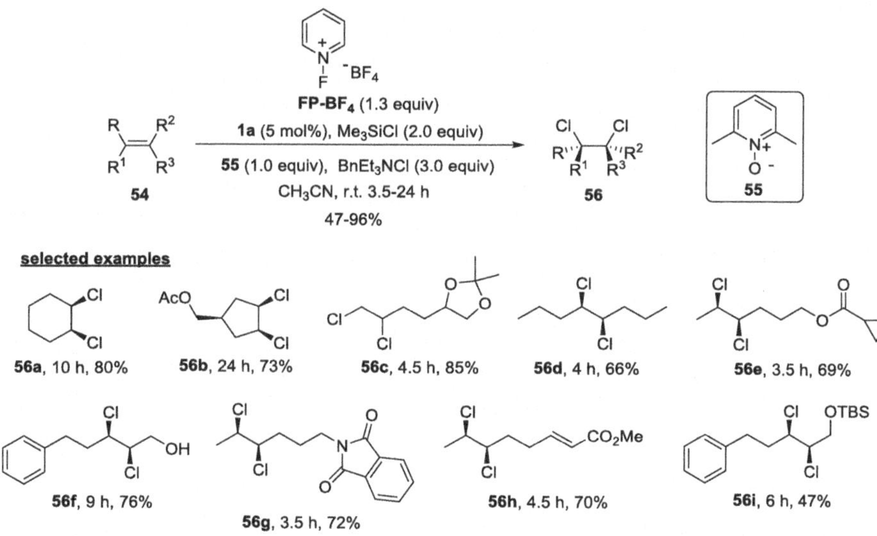

Scheme 1.26 Se-catalyzed *syn*-stereospecific dichlorination of alkenes

syn-1,2-dichlorides **56** from cyclic and acyclic 1,2-disubstituted alkenes **54** (Scheme 1.26). The presence of 2,6-lutidine *N*-oxide **55** (1.0 equiv.) as an additive accelerates the reaction and does not reduces the diastereoselectivity. In the optimization studies, the authors used ^{n}Bu$_4$NCl as chloride source with excellent results; however, less expensive BnEt$_3$NCl presented the same performance, being chosen instead. Selectfluor® could also be used as stoichiometric oxidant in the place of PF-BF$_4$, albeit yields were slightly inferior. An over stoichiometric amount of Me$_3$SiCl (2.0 equiv.) was used to suppress the fluoride ion generated from the oxidant, increasing thus the reaction yield by the formation of unreactive M$_3$SiF and releasing more chloride ion in the reaction media. Differently substituted diaryl diselenides were tested as pre-catalysts and only the electron-rich bis(4-methoxyphenyl)diselenide showed similar results than **1a**. However, because in some cases the 4-methoxyphenyl derivative increased the amount of elimination products, **1a** was chosen for the reaction.

In the proposed mechanism for the *syn*-dichlorination, the authors suggested that (PhSe)$_2$ **1a** is firstly oxidized by PF-BF$_4$ in the presence of chloride ion to PhSeCl$_3$, the active catalyst in the reaction (Scheme 1.27). The addition of PhSeCl$_3$ to alkene **54** could involve the loss of a chloride to form the electrophilic species PhSeCl$_2$$^+$ **A** that, after addition to **54** forms the seleniranium ion **B**. A nucleophilic ring-opening by a chloride ion gives the β-chlorinated selenide **C**, followed by an invertive displacement of the Se(IV) group with chloride to give the *syn*-dichloride **56** and PhSeCl. The dissociation of one chloride ligand from Se before substitution in **C** could increase the nucleofugality of the selenium group by rendering it positively charged. The stereochemistry of the deselenylation step by a second chloride ion is attributed to the fact that the substituents near the chloro did not anchimerically assist the reaction.

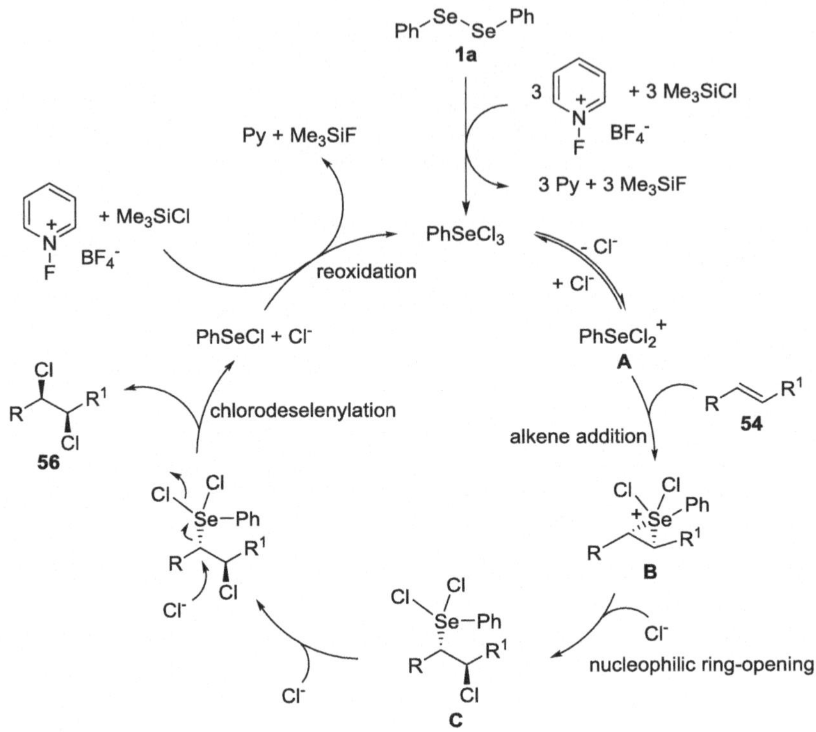

Scheme 1.27 Proposed mechanism for the *syn*-dichlorination of alkenes

1.3 Se-Catalytic Activation of Halogenating Agent

The Se-catalyzed oxidative chlorination of alkenes, firstly described by Sharpless in 1979, was revisited by Tunge and coworkers in 2004 [59], who used a catalytic amount of PhSeCl to activate *N*-chlorosuccinimide (NCS) in the allylic chlorination of β,γ-unsaturated esters **2**, acids **34**, nitrile **57** and allylbenzene **58** (Scheme 1.28). The allyl chlorides **59a–h** were obtained in 62–89% yields after 4–48 h of reaction at 25 °C using 5 mol% of PhSeCl and 1.1 equiv. of NCS. The use of higher amounts of NCS or the absence of molecular sieves negatively influenced the reaction rates. Longer reaction times and a larger amount of catalyst (20 mol%) were required for nitrile **57** and allylbenzene **58**, while the unsaturated esters **2a–c** were more reactive (4–8 h) compared to the acids analogues **34** (16 h of reaction). When prenyl olefins **60** were used, the respective allyl chlorides **61** were selectively formed in good yields. The reactions were faster than those with β,γ-unsaturated acids **34**, except for prenol **60b**, that required 48 h at 35 °C to deliver the respective chlorohydrin **61b** in 68% yield (Scheme 1.28).

A proposed catalytic cycle for the formation of allyl chloride **59a** is showed in Scheme 1.29. Initially, the reversible 1,2-addition of PhSeCl to olefin **2a** occurs to

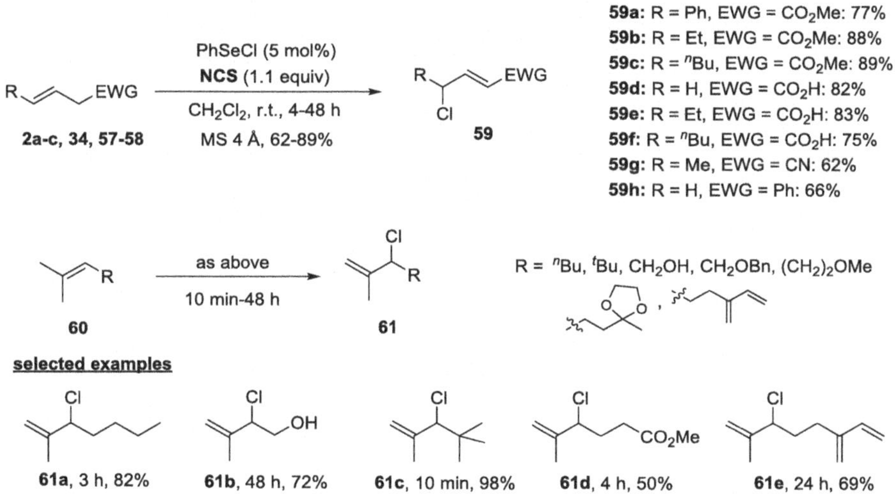

Scheme 1.28 Se-catalyzed oxidative chlorination of alkenes using PhSeCl/NCS

form the alkyl selenide **A**. Then, intermediate **A** is oxidized by NCS to the unstable tetravalent selenenyl chloride **B** that, after elimination of succinimide and allyl chloride **59a** affords PhSeCl, completing the catalytic cycle. The inhibition of the reaction by NCS could be explained by the formation of the inactive Se(IV) complex **62**, which is in equilibrium with catalyst PhSeCl. The positive effect of adding molecular sieves could be attributed to the protection of intermediates **62** and **B** from hydrolysis.

The involvement of PhSe(succinimide)Cl_2 **62**, formed by the oxidative addition of NCS to PhSeCl, in the allylic chlorination described in Scheme 1.28, motivated Tunge to investigate whether this complex could catalyze the α-chlorination of ketones [73]. The fact that **62** is less active in the allylic chlorination allowed the selective α-chlorination even in the presence of double bonds. Several monochlorinated products **64** were obtained from the respective ketones **63** in good to very good yields (61–95%) after 10 min to 48 h at room temperature, using PhSeCl (5 mol%) and NCS (1.1 equiv.) in acetonitrile as the solvent (Scheme 1.30). The authors observed that ebselen (5 mol%) could also be used as a catalyst in the synthesis of α-chloroketone **64a**, however a longer reaction time was necessary, due to its poor solubility in acetonitrile [74].

When cyclohexanone **63f** was subjected to the reaction conditions in CH_3CN, the selectivity was lost, and a 4:1 mixture of monochlorinated and dichlorinated cyclohexanone was obtained. This problem was circumvented using methanol as solvent and α-chloroacetophenone **64f** was the only product (18 h, 72% yield). In this case, the monochlorinated product was trapped with methanol, forming the acetal **A**, which delivered α-chlorocyclohexanone **64f** after filtration using silica (Scheme 1.31) [73]. By using PhSeBr/NBS in the place of PhSeCl/NCS, α-bromocyclohexanone was obtained in 86% yield. Finally, when α,β-unsaturated

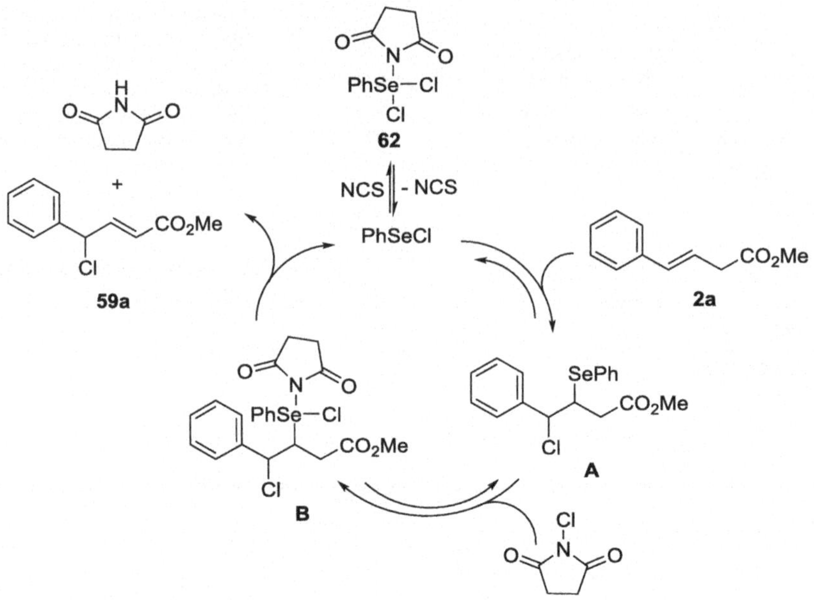

Scheme 1.29 Proposed catalytic cycle for the selenochlorination-deselnylation using PhSeCl/NCS

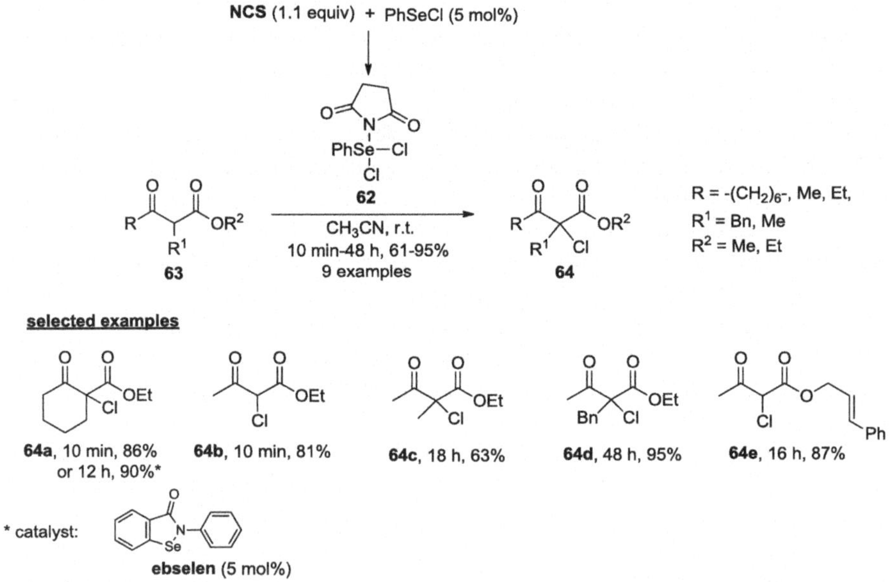

Scheme 1.30 Se-catalyzed α-chlorination of ketones using PhSeCl/NCS

ketone, mesityl oxide **63g**, reacted with NCS/PhSeCl in CH_3CN, vinyl chloride **65g** was isolated in 79% yield after 7 h. A reverse selectivity, however, was observed replacing the solvent for MeOH, giving exclusively the methyl chlorinated product **64g** (72% yield).

An investigation on the possible intermediates involved in the monochlorination of β-ketoesters showed that the mechanism most likely involves an electrophilic chlorination rather than a selenylation [73]. This indicates that in some way, PhSeCl enhances the electrophilicity of the halogen source (NCS in this case), through the formation of the Se(IV) complex **62**. The chloronium ion formation could not be involved in the formation of vinyl chloride **65g**, once PhSeCl addition to the double bond followed by oxidation with NCS would also afford **65g**.

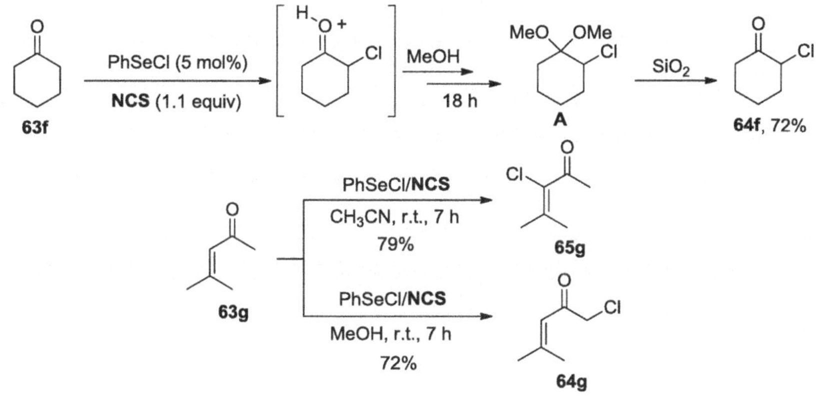

Scheme 1.31 Selectivity in the Se-catalyzed monochlorination of ketones

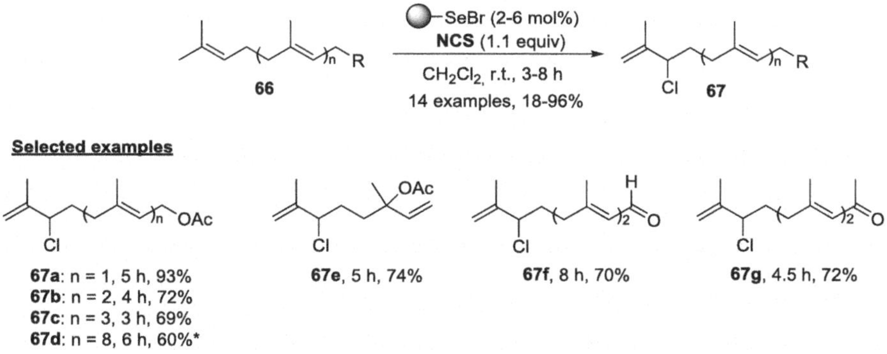

* A 3:1 mixture with internal derivatives.

Scheme 1.32 Allyl chlorides from polyprenoids using solid-supported selenenyl bromide (P-SeBr) as catalyst

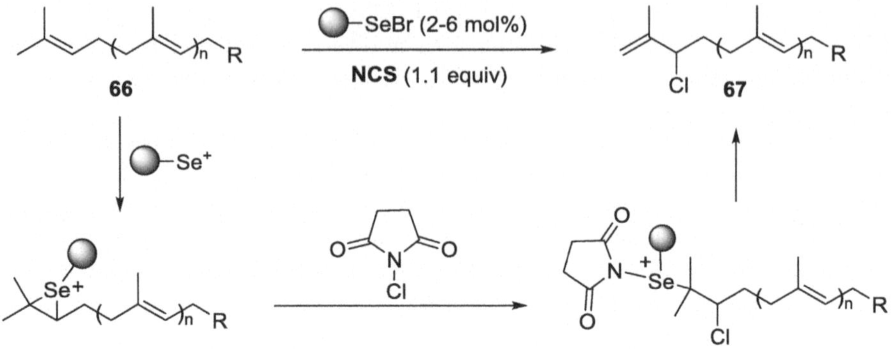

Scheme 1.33 Proposed mechanism of the monochlorination catalyzed by P-SeBr

A complementary protocol to selectively prepare allyl chlorides from polyprenoids was developed by Barrero and coworkers using a catalytic amount (2–6 mol%) of solid-supported selenenyl bromide (P-SeBr) and NCS (1.1 equiv.) as the chlorinating agent [75]. A diversity of polyunsaturated isoprenoids **66** was efficiently converted to the respective allyl chlorides **67** in 18–96% yield after 3–8 h of reaction at room temperature (Scheme 1.32).

The authors credited the high selectivity of the reaction to the relatively big size of the electrophilic selenium catalyst (polymer-supported Se^+) combined with its very low concentration, that could lead to the preference for the terminal isopropylidene group in the polyolefinic polyprenoids (Scheme 1.33).

1.4 Organoselenium as Lewis Base Catalysts

An emergent facet of selenium compounds in catalysis is in Lewis base catalysis, i.e., in the activation of Lewis acids [76]. The adduct of the reaction of an organoselenium Lewis base (**LB**) donor with a Lewis acid (**LA**) acceptor is a three-center four-electron hypervalent bonded cationic Lewis acid **LA**$^+$ (Scheme 1.34). The cationic species **LA**$^+$ possess an increased electrophilicity compared to the Lewis acid precursor.

The first report on the use of organoselenium as Lewis base catalyst was authored by Tunge and coworkers, who described the Se-catalyzed bromolactonization of unsaturated carboxylic acids **34** and **45** using $(PhSe)_2$ **1a** as a Lewis base catalyst and NBS as brominating agent [77]. Several experimental evidences led the authors to describe, for the first time, the catalytic Lewis base activation of *N*-bromosuccinimide (NBS) by **1a** in the selective synthesis of β and γ-lactones **68** and **69** (Scheme 1.35). By stirring a mixture of the unsaturated acid with NBS (1.1 equiv.) in the presence of **1a** (5 mol%) in acetonitrile at −30 °C for 1–4 h, ten differently substituted bromolactones were obtained in 33–90% yields. The selectivity for the γ-lactones over the β-lactones proved to be due a kinetic control

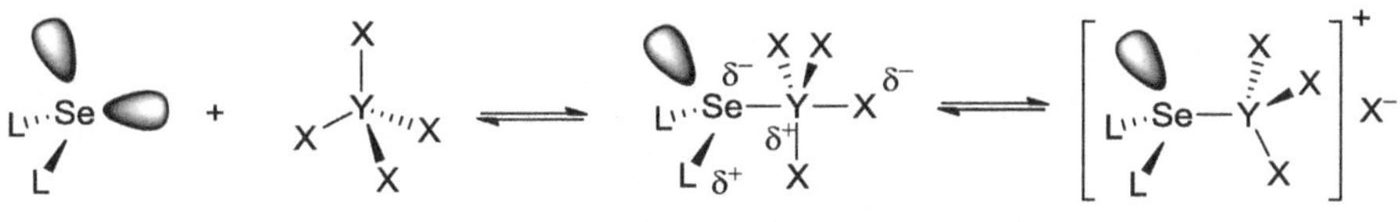

Scheme 1.34 Lewis base catalysis using organoselenium compounds

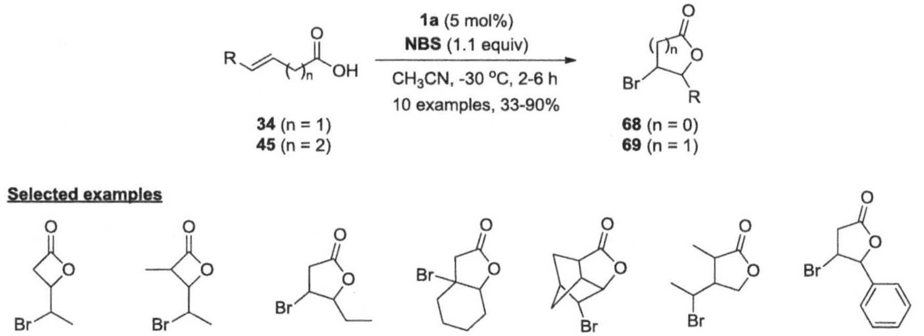

Scheme 1.35 Se-Catalyzed bromolactonization of unsaturated carboxylic acids using NBS

rather than a thermodynamic one, once no equilibration of a 2:1 mixture of β-lactone and the γ-lactone was observed under the reaction conditions.

In the proposed mechanism, the authors suggest that the first step is the nucleophilic attack of $(PhSe)_2$ **1a** to NBS, producing the cationic selenium complex **LA**$^+$, with a succinimide counter ion (Scheme 1.36). The succinimide anion deprotonates carboxylic acid **34** giving the respective carboxylate. The observed regiocontrol in the formation of the butyrolactone **69** was rationalized as involving a nucleophilic displacement of a selenium bromonium ion **A**.

A close related reaction was developed by Seoane and coworkers in the Se-catalyzed synthesis of iodohydrins from alkenes [78]. The authors used *N*-iodosuccinimide (NIS) as iodine source and $(PhSe)_2$ **1a** (5 mol%) as the catalyst in the presence of equivalent amounts of water. The respective iodohydrins **70** were obtained in 40–81% yields after 1–10 h at room temperature (Scheme 1.37). The mechanism of the reaction is most likely similar to that of Scheme 1.36, for the bromolactonization, except that water is the nucleophile in the ring opening of the selenium iodonium intermediate, delivering the expected iodohydrin **70** and regenerating the catalyst **1a** for a new cycle.

Leung and Yeung developed a selenium Lewis base-catalyzed chloroamidation of olefins using Ph_2Se **71a** (20 mol%) as the catalyst, NCS as chlorine source and acetonitrile as nucleophile using a mixture of CH_3CN and H_2O as the solvent [79]. Starting from cyclohexene, the authors performed a study on the best conditions for the reaction (catalyst, amount of water, halogen source, time) to prepare haloamides. Different Lewis bases were tested, including triphenylphosphine oxide, sulfide and

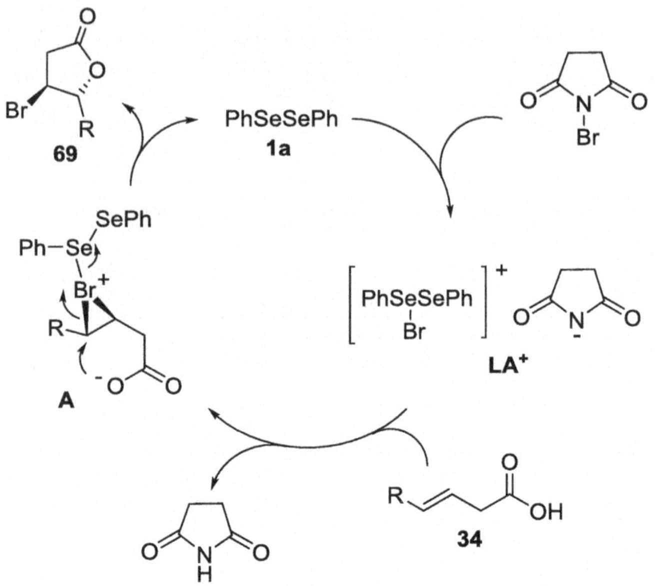

Scheme 1.36 Proposed mechanism of the Se-catalyzed bromolactonization

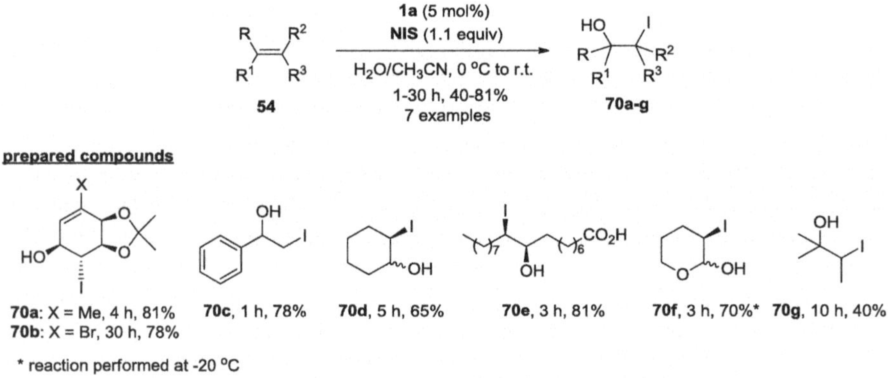

Scheme 1.37 Se-Catalyzed synthesis of iodohydrins from alkenes using NIS

selenide, thioureas and selenoureas, $(Ph)_2S$, $(Ph)_2Se$, and $(PhSe)_2$ **1a** and the best result was obtained using 20 mol% of diphenyl selenide in the presence of 1.2 equiv. of water. Regarding the halogenating agent, the use of 2 equiv. of NCS delivered the expected chloroamide in higher yield after 24 h of reaction. The Markovnikov adducts were preferentially obtained in all the examples but using triphenyl(vinyl) silane, that afforded the anti-Markovnikov adduct **72g** in 86% yield after 18 h of reaction (Scheme 1.38).

It is noteworthy that in the previously described reaction (Scheme 1.37), using 5 mol% of $(PhSe)_2$ **1a** as catalyst and NIS as halogenating agent in the same solvent

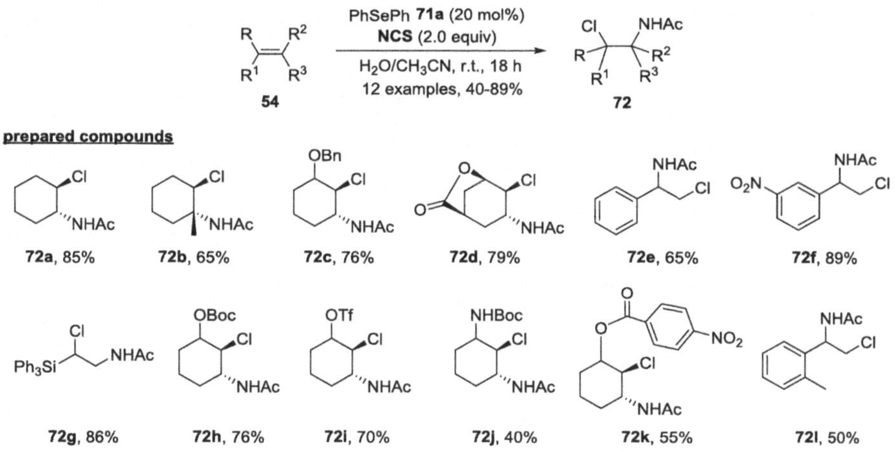

Scheme 1.38 Chloroamidation of olefins using Ph_2Se as the catalyst

(CH_3CN/H_2O), water was the nucleophile instead CH_3CN. Under the PhSePh Lewis base catalysis, NIS did not react, even after 24 h of stirring at room temperature. The method was successfully extended to the reaction of cyclohexene with propionitrile and benzonitrile, affording *N*-(2-chlorocyclohexyl)-propionamide and *N*-(2-chlorocyclohexyl)benzamide in 70% and 83% yields after 48 h of reaction at room temperature.

The proposed reaction mechanism involves in the first step the Lewis base activation of NCS by PhSePh **71a**, to form the 3c–4e (3 centers–4 electrons) complex intermediate **A**, which is in equilibrium with the selenium cationic species **B** and succinimide anion (Scheme 1.39). Then, the highly electrophilic chlorine in **B** reacts with olefin **54a** to form the chloronium ion intermediate **C**. Following, a nucleophilic attack by CH_3CN regenerates selenide **71a** for a new catalytic cycle, yielding intermediate **D**, which is hydrolyzed to chloroamide **72a**.

Electron-rich bis(4-methoxyphenyl)selenide **71b** was used by Zhao and coworkers in the Lewis base activation of *N*-SCF_3 bonds, allowing the insertion of the -SCF_3 group instead -Cl in terminal and internal alkenes and a concomitant amination, giving excellent yields of trifluoromethylthio amides **74** [80]. The authors started the optimization studies by using styrene **54e** as the alkene, diphenyl selenide **71a** as the catalyst and different sources of -SCF_3 using acetonitrile as the solvent. However, only trace amounts of the desired amide were observed after 24 h of reaction. The best results were obtained when 10 mol% of **71b** was used as the catalyst and 1.3 equiv. of *N*-trifluoromethylthiosaccharin (NTFSSac) was the -SCF_3 source, in the presence of TfOH (1.0 equiv.) as a Brønsted acid, H_2O (1.2 equiv.), CH_3CN (20 equiv.) and hexane as the solvent, for 24 h at room temperature. This optimal condition was extended to the reaction of styrene **54e** with differently substituted nitriles **73** and good to excellent yields of **74** were obtained using alkyl and aryl-substituted nitriles (Scheme 1.40).

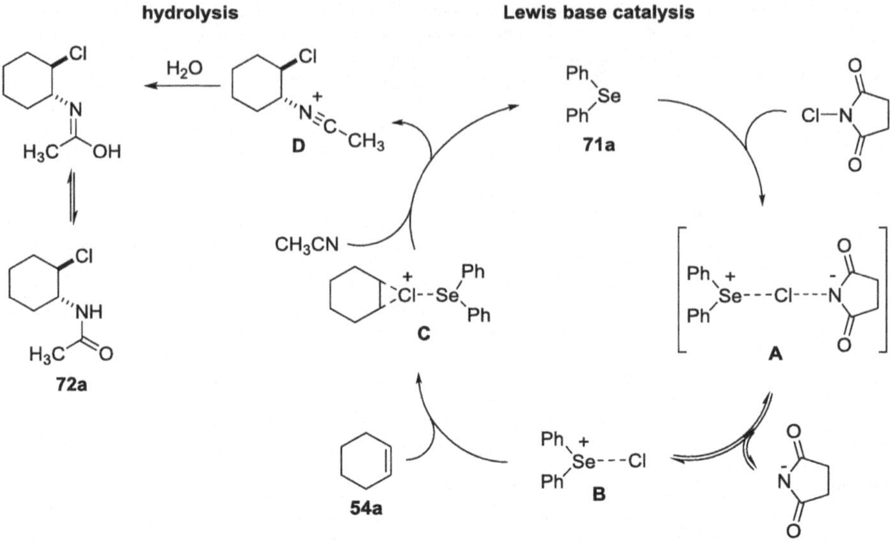

Scheme 1.39 Two steps mechanism proposed for the Se-catalyzed chloroamidation of olefins

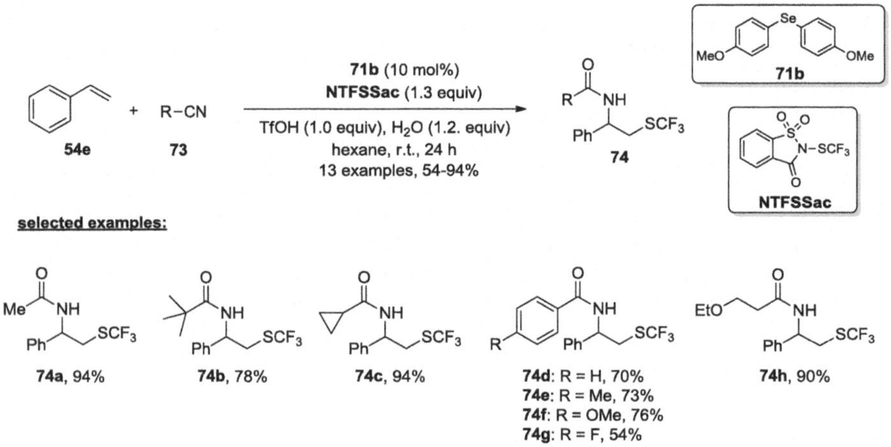

Scheme 1.40 Se-catalyzed synthesis of trifluoromethylthio amides from styrene

Using acetonitrile **73a** as the solvent, the protocol was suitable to promote the trifluoromethylthioamidation of differently substituted alkenes **54**, including internal ones (Scheme 1.41). (*E*)-Oct-4-ene (*E*)-**54d** reacted with NTFSSac in the presence of TfOH/H_2O to afford exclusively the *cis*-amide **74q** in 79% yield, while (Z)-oct-4-ene (Z)-**54d** delivered *trans*-amide **74q** as the only isomer in similar yield (Scheme 1.41). These outcomes revealed the high level of stereoselectivity of the reaction. Changing acetonitrile for acetic acid as the solvent, the authors were able to perform the trifluoromethylthioesterification of several alkenes, giving the -SCF_3

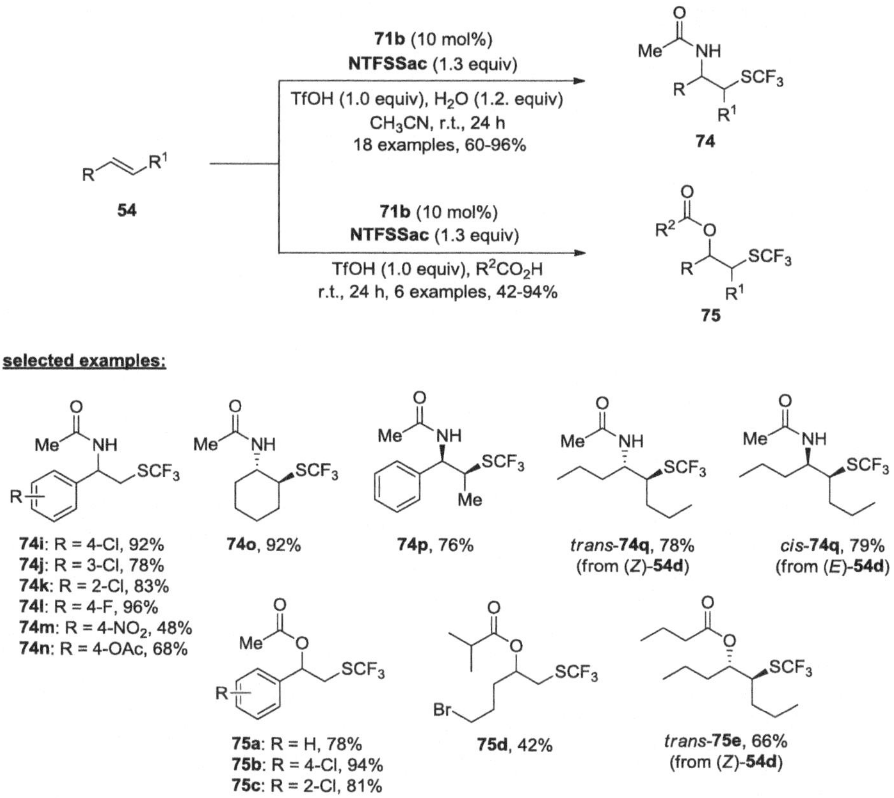

Scheme 1.41 Se-catalyzed trifluoromethylthioamidation of alkenes using MeCN as the solvent

functionalized esters **75** in very good yields. The protocol was suitable to other carboxylic acids, with good results and selectivity for internal alkenes (Scheme 1.41).

As showed in Scheme 1.42, for the reaction of cyclohexene **54a** with NTFSSac and acetonitrile, the proposed mechanism is close-related to the chloroamination with NCS, except that -SCF_3 is coordinated with the alkene instead Cl. The role of triflic acid in the reaction is to protonate the saccharin, favoring the formation of the highly reactive selenium cation **B** from the hypervalent complex **A**. In the presence of the alkene **54a**, the trifluoromethylthiiranium ion **C** is formed, which reacts with acetonitrile to form **D** by an *anti*-addition, opening the iranium ring. After hydrolysis, the desired functionalized amide **74o** is formed.

Bis(4-methoxyphenyl)selenide **71b** was used by the same authors in a combined Lewis base catalysis and redox chemistry in the synthesis of SCF_3-containing tertiary alcohols from alkenes [81]. Alkenes functionalized with carboxyl, hydroxy, sulfonamide, or ester groups are smoothly converted to trifluoromethylthiolated *O*- and *N*-heterocycles. In a typical procedure, a mixture of alkene **54**, NTFSSac (1.3 equiv.), H_2O (5.0 equiv.), and **71b** (10 mol%) under oxygen atmosphere in $MeNO_2$ as the solvent is stirred at room temperature for 24 h, affording the desired

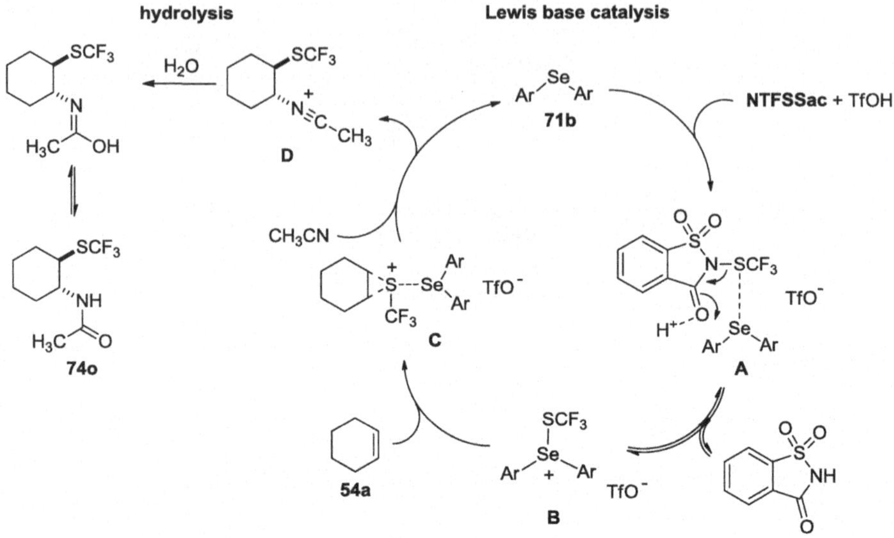

Scheme 1.42 Proposed mechanism for the Se-catalyzed trifluoromethylthioamidation of alkenes

SCF_3-functionalized alcohols **76** or heterocycles **77** in 38–98% yields (Scheme 1.43). Several alkyl and aryl alkenes were selectively functionalized, with the poorest results being observed when electron-rich aryl groups were present in the substrates. The authors have observed that increasing or reducing the electronic density in the diaryl selenide led to lower yields of product. Besides, the presence of $MeNO_2$ as a solvent, as well as the O_2 atmosphere was crucial to afford good results. Differently from the synthesis of SCF_3-containing amides **74** and esters **75** described on Schemes 1.40 and 1.41, the presence of Brønsted acid TfOH in this reaction is prejudicial, once it decreases the nucleophilicity of the weak nucleophile H_2O and promotes a competitive elimination reaction in the starting alkene, forming allylic compounds instead the desired functionalized alcohols **76**, mainly when alkyl alkenes are used.

Several control experiments were performed to elucidate the role of oxygen in the reaction, including labeling experiments using $^{18}O_2$ and $H_2{}^{18}O$. From these experiments it came clear that water is responsible for the origin of the hydroxyl in the the alcohol **76**. The authors have prepared the selenoxide **A**, derivative from **71b**, and used it as the catalyst in the reaction of styrene **54e** with NTFSSac under N_2 atmosphere. The expected alcohol **76a** was obtained in 52% yield, even in the absence of O_2, a slightly better result compared to the observed when **71b** was the catalyst (only 17% yield after 24 h). Based on these and other experiments, a plausible mechanism was proposed that involves a combined Lewis base catalysis by **71b**, activating the electrophile NTFSSac and a redox reaction of Se(II) to Se(IV). This redox process is responsible by generation in situ of TfOH, which aids the elimination of saccharin from NTFSSac, as previously discussed. In the redox step, selenide **71b** is firstly oxidized to selenoxide **A** by O_2/$MeNO_2$. Following, **A** is

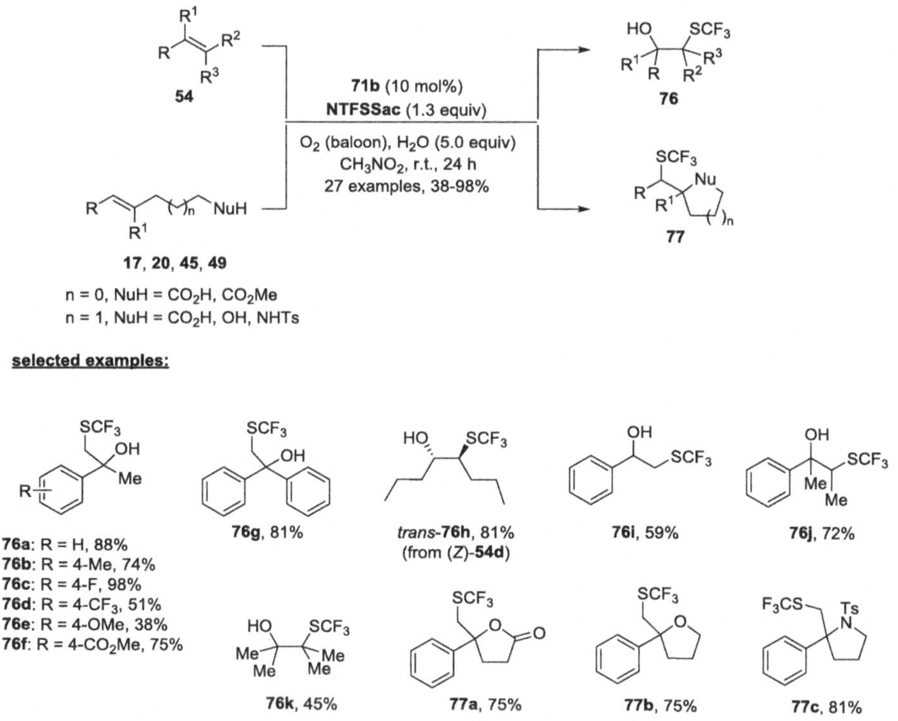

Scheme 1.43 Se-catalyzed synthesis of SCF_3-containing tertiary alcohols and heterocycles from alkenes

reduced to **71b** by NTFSSac in the presence of water, along with the releasing of TfOH. Together, TfOH and **71b** activate NTFSSac, giving intermediate **B**, that reacts with alkene **54** to form the episulfonium ion **C**. Reaction between **C** and water leads to the functionalized alcohol **76** and TfOH, regenerating **71b** for a new cycle (Scheme 1.44).

Kumar and coworkers prepared several ebselen-like isoselenazolones **78** and tested them as Lewis bases in the activation of brominating agents (Br_2, NBS) in the bromolactonization of unsaturated acids, bromoesterification of alkenes and oxidation of alcohols [82]. Among the tested Se-catalysts, the electron-deficient 5,6-difluoro-2-(1-phenylethyl)benzo-[*d*][1,2]selenazol-3(2*H*)-one **78a** was the more active, affording good yields and selectivity in most of the reactions. In the bromolactonization of γ,δ-unsaturated carboxylic acids **45**, the authors observed that molecular bromine (1.2 equiv.) in the presence or not of a base (K_2CO_3; 1.5 equiv.) was a better bromine source than NBS, using **78a** (1 mol%) as the catalyst. The expected brominated γ-butyrolactones **69** were obtained in 86–95% yields after 3–12 h at room temperature (Scheme 1.45). In all the examples, NBS successfully replaced Br_2 as a source of Br^+, with similar yields and selectivity, without need of K_2CO_3. However, the difficult in separating the bromolactones **69** and **79** from succinimide generated in the reaction made molecular bromine to be chosen over NBS.

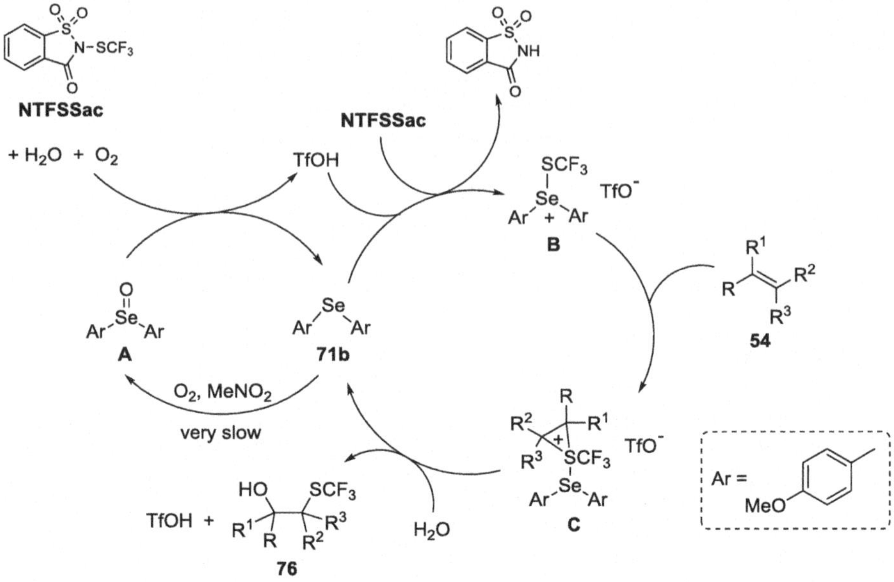

Scheme 1.44 Proposed mechanism for the Se-catalyzed trifluoromethylhydroxylation

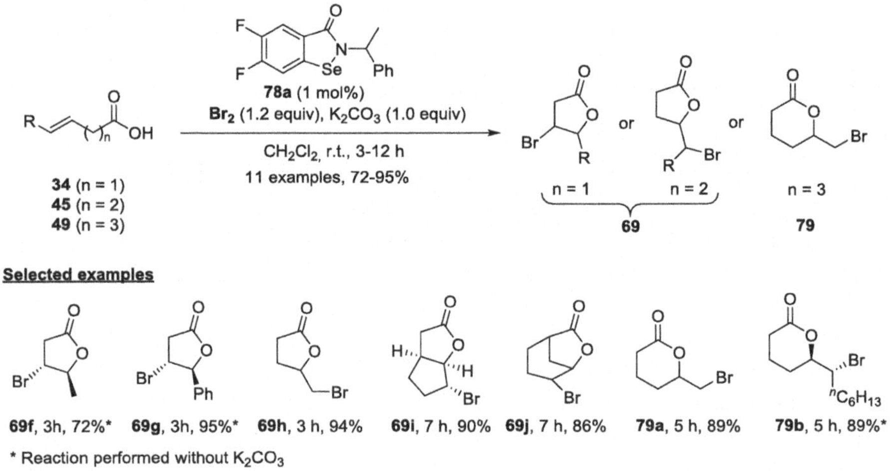

Scheme 1.45 Se-catalyzed bromolactonization of unsaturated acids, bromoesterification of alkenes and oxidation of alcohols using Br_2 and NBS

In contrast to the bromolactonization of unsaturated acids **34**, **45**, and **49**, which is entropically favored when five- and six-membered ring are formed, the intermolecular bromoesterification of non-activated, simple alkenes **54** is not trivial. For an example, the bromoesterification of cyclooctene with benzoic acid **80a** (1.0 equiv.) and NBS (1.5 equiv.) afforded less than 10% of the expected bromoester **81a** after

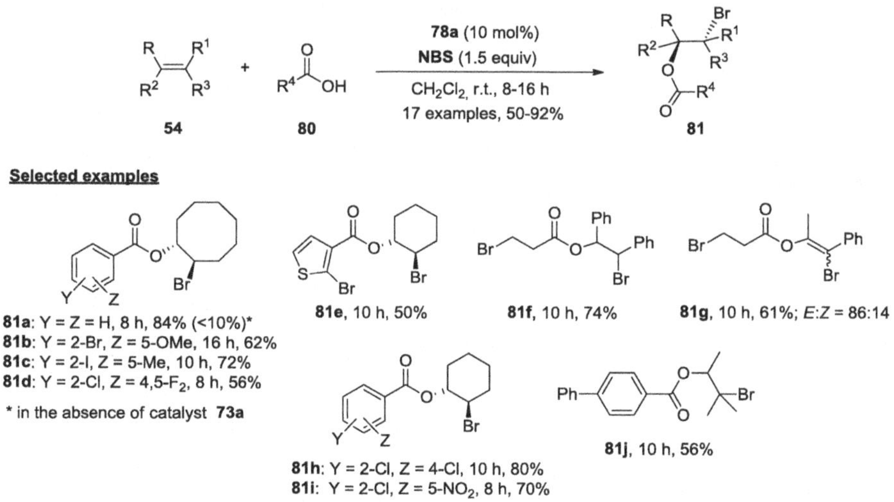

Scheme 1.46 Se-catalyzed bromoesterification of non-activated alkenes

8 h. If 10 mol% of isoselenazolone **78a** is added, however, a positive Lewis base catalytic effect is observed, and bromoester **81a** could be isolated in 84% yield after 8 h [82]. This Se-catalyzed bromoesterification was extended to the reaction of other differently substituted alkenes and carboxylic acids, with good results also for 1-phenylpropyne, that delivered, after 10 h of reaction with 2-bromopropanoic acid and NBS, 61% of the bromovinyl ester **81g** in a *E*:*Z* ratio of 86:14 (Scheme 1.46).

A third application of isoselenazolone **78a** was showed in the Se-catalyzed oxidation of secondary alcohols to ketones by molecular bromine [82]. The catalytic effect of **78a** can be verified in the reaction of 1-phenylethyl alcohol **82a**, that did not react with Br_2 (1.5 equiv.) after 3 h at room temperature. When 10 mol% of **78a** was added, complete oxidation of alcohol **82a** was observed in 30 min, giving acetophenone **83a** in 90% yield. The protocol was extended to several benzyl alcohol derivatives, affording the respective ketones in 60–84% yields after 14–40 h. To avoid undesired side products, the reaction temperature was around −35 to −20 °C in most of the reactions (Scheme 1.47). When cyclohexanol was the substrate, cyclohexanone **83k** was obtained in 44% yield after 16 h at −20 °C.

Supported by ^{77}Se-NMR and MS analysis of the intermediates involved in the reaction, the authors proposed a mechanism for both, the bromolactonization and the alcohol oxidation reactions catalyzed by isoselenazolone **78** (Scheme 1.48) [82]. The key step of the reaction is the formation of the complex **A** (or **B**), from the Lewis base activation of Br_2 (or NBS) by **78a**. By ^{1}H-NMR analysis, the authors suggested that complex **A** is in equilibrium with its ionic form **C** (broad signals were observed in the NMR spectrum). The reaction of β,γ-unsaturated carboxylic acids **45** with complex **A** or **B** in the presence of K_2CO_3 leads to the bromolactone **69** and regenerates the catalyst **78a** for a new cycle. In this reaction, there is the formation of HBr or succinimide as co-products. The mechanism for the Se-catalyzed

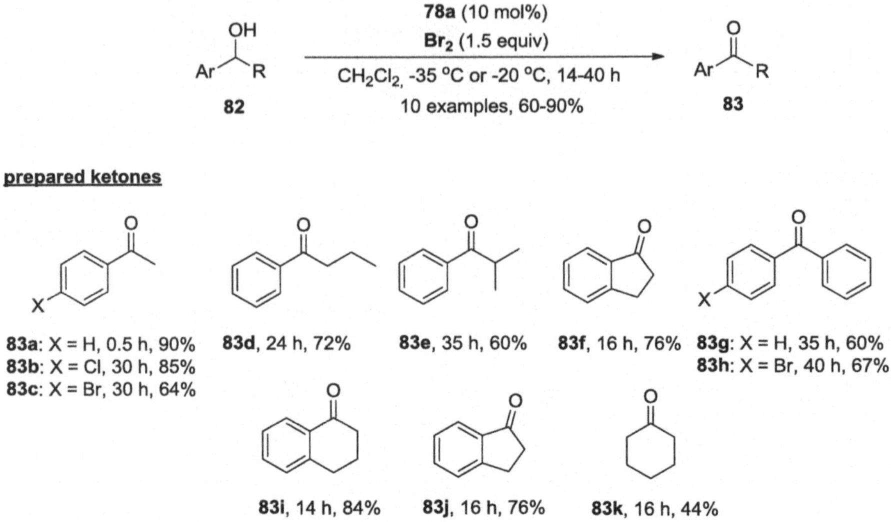

Scheme 1.47 Se-Catalyzed oxidation of secondary alcohols to ketones by Br_2

bromine-promoted oxidation of secondary alcohols is quite similar, with the alcohol the reaction occurs via the formation of the Swern-like alcoxyselenonium ion **G**. Once formed, **G**, that is in equilibrium with **F**, loses a proton to give the arylacetophenone **83** and HBr, regenerating the catalyst **78a**.

More recently, a new optically active diallyl selenide, (α-pinene)$_2$Se **71c** was prepared and used by Houssame, Santi, and coworkers as a Lewis base catalyst in the allylic chlorination of terpenes [83]. The new selenide was obtained in 71% yield by a new protocol, starting from β-pinene and SeO_2 in the presence Et_3N. Catalyst **71c** (5 mol%) efficiently activated NCS (1.1 equiv.) in the reaction with several terpenoids **84**, affording the respective allyl chlorides **83** in good yields (73–92%) and excellent selectivity after 3–5 h at room temperature (Scheme 1.49).

Denmark, in its studies on Lewis base activation of Lewis acids developed a new class of organoselenium Lewis base analogue to HMPA, the selenophosphoramide (Se-HMPA) **86a**, which successfully activates *N*-phenylselenosuccinimide (NPSS) in the selenolactonization reaction of (*E*)-4-phenyl-but-3-enoic acid (*E*)-**34e** [84]. In the absence of catalyst, only 8% of product was obtained after 3 h. The catalytic activity of Se-HMPA **86a** was higher than HMPA and the thiophosphoramide analogue (S-HMPA). While HMPA afforded the expected phenylseleno-lactone **87** *trans* in 60% after 3 h, S-HMPA (10 mol%) delivered 89% after 10 min and Se-HMPA afforded **87** in 95% yield after only 5 min of reaction. The reaction is stereospecific, and (*Z*)-4-phenyl-but-3-enoic acid (*Z*)-**34e** reacted with NPSS (1.2 equiv.) in the presence of **86a** (10 mol%) to give exclusively the *cis*-lactone **87** in 94% after 3 h at room temperature (Scheme 1.50).

The proposed catalytic cycle of the Lewis base catalyzed selenolactonization starts by the coordination of Se-HMPA **86a** with NPSS, to form the hypervalent

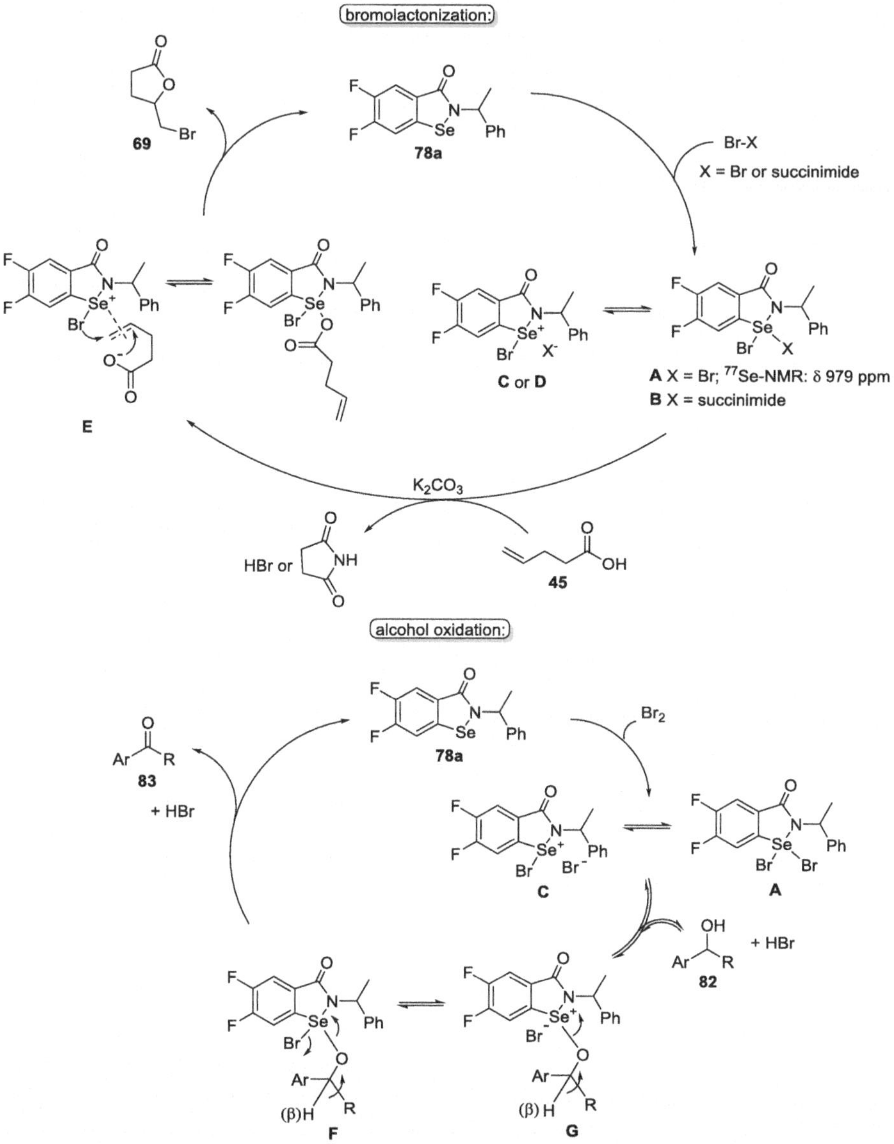

Scheme 1.48 Proposed mechanisms of the bromolactonization and the alcohol oxidation reactions catalyzed by isoselenazolone **78**

Lewis base-NPSS complex **A** (Scheme 1.51). Then, protonation of succinimide by unsaturated acid **34e** affords cationic selenium species **B**, which reacts with the double bond of carboxylate to give seleniranium **C**. After intramolecular attack and ring opening of the seleniranium by the pendant carboxylate in **C**, the Lewis base **86a** is eliminated, together the formation of the selenolactone **87**.

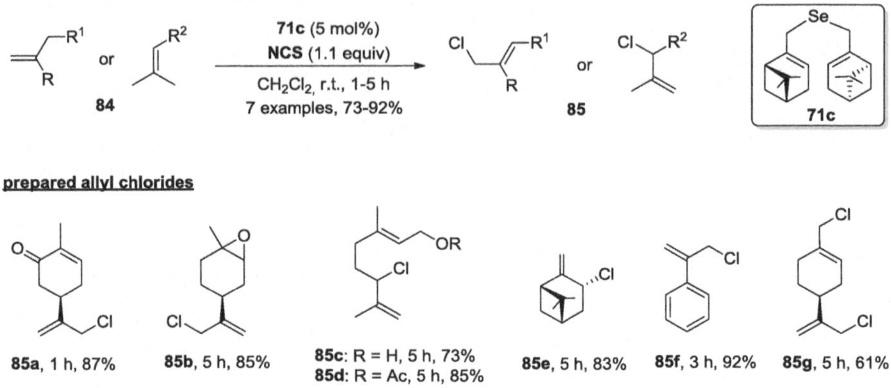

Scheme 1.49 Allylic chlorination of terpenes catalyzed by (α-pinene)$_2$Se

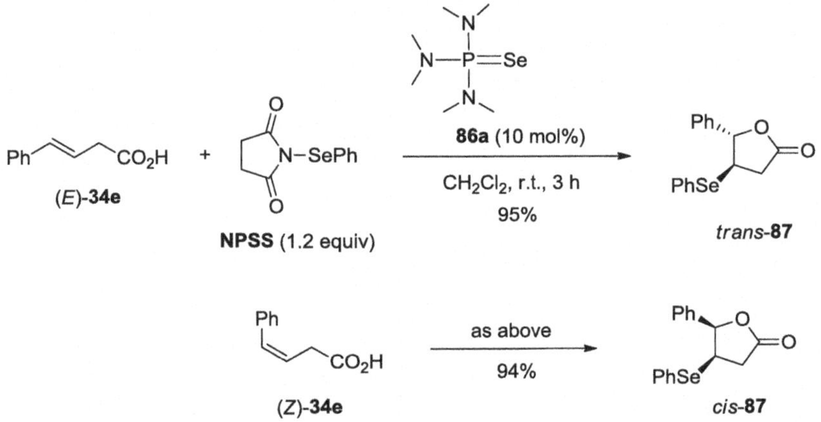

Scheme 1.50 Selenophosphoramide (Se-HMPA) as a catalyst in selenolactonization reactions

After the good results using Se-HMPA **86a** in the Lewis-base activation of electrophilic selenium species in selenofunctionalization reactions, Denmark moved its focus to the development of chiral Se-Lewis base able to catalyze the thiofunctionalization of unactivated alkenes [85]. The chiral (*R*)-BINAM-based selenophosphoramide **86b** efficiently promoted the activation of *N*-phenylsulfenylphthalimide (NPSP) in the reaction with alkenes **88** in the ring-closure with the pendant hydroxyl group. Basically, the reaction consists in stirring a mixture of alkene **86**, NPSP (1.0 equiv.), MsOH (1.0 equiv.), and the catalyst **86a** (10 mol%) in CH_2Cl_2 (0.4 M) at −10 or −20 °C for 48 h. The low temperature is important to increase the enantioselectivity, while the Brønsted acid helps in the elimination of phthalimide from the NPSP to form the active electrophile. The method was applied to a diversity of simple alkenes and styrenes, affording the respective sulfenylated tetrahydropyrans (*S*-THP) **89** or tetrahydrofurans (*S*-THF) **90** in

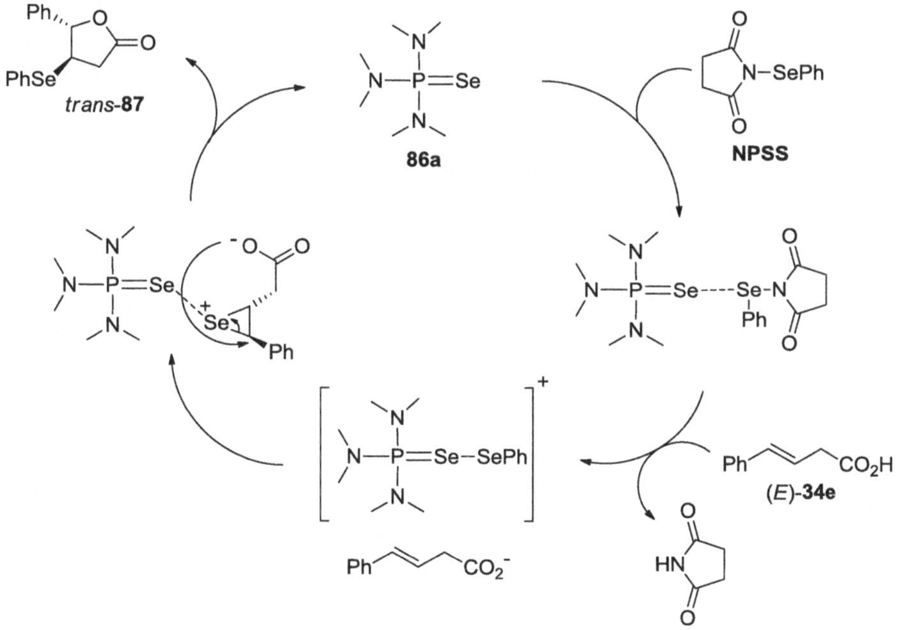

Scheme 1.51 Proposed mechanism for the Se-HMPA-catalyzed selenocyclization

24–96% yields and up 92% *ee* (Scheme 1.52). Electron-poor alkene **88b** showed to be less reactive, affording the expected S-THP **89b** in only 41% yield. In contrast, electron-rich alkenes were as reactive as the unsubstituted styryl derivative **88a**. Trisubstituted alkene **88h** reacted slowly, and the reaction was conducted at room temperature, affording many side products and low enantioselectivity, with only 24% of the respective S-THP **89h**.

When (*E*)-5-phenylpent-4-enoic acid **45b** was used, the γ-butyrolactone **91** was obtained in 83% yield and 82% *ee* after 48 h at −10 °C. The method was successfully used with exogenous nucleophiles, as showed for the intermolecular capture of the thiiranium intermediate generated from internal and terminal alkenes **54** and NPSP in the presence of (*R*)-**86b**, affording good yields and excellent selectivity of the respective thioethers **92** and **92′** (Scheme 1.53).

The authors used ^{31}P-NMR to support a mechanistic proposal for the Se-catalyzed enantioselective thiofunctionalization of alkenes **54** or **88** (Scheme 1.54). The first step of the catalytic cycle is the formation of the hypervalent intermediate **A**, by the coordination of the Lewis base (*R*)-**86b** with -SPh, mediated by the Brønsted acid MsOH. The involvement of **A** was evidenced by disappearance of the signal at $\delta = 91.6$ ppm characteristic of (*R*)-**86b** and the presence of a new signal at 60.4 ppm in the ^{31}P-NMR spectrum, attributed to complex **A**. In contact with alkene **54** or **88**, the sulfenium ion is transferred from **A** to the double bond, forming the thiiranium ion **B**, which represents the enantiodetermining step, controlled by the architecture of the catalyst (*R*)-**86b**. Then, an intra or intermolecular nucleophilic attack occurs

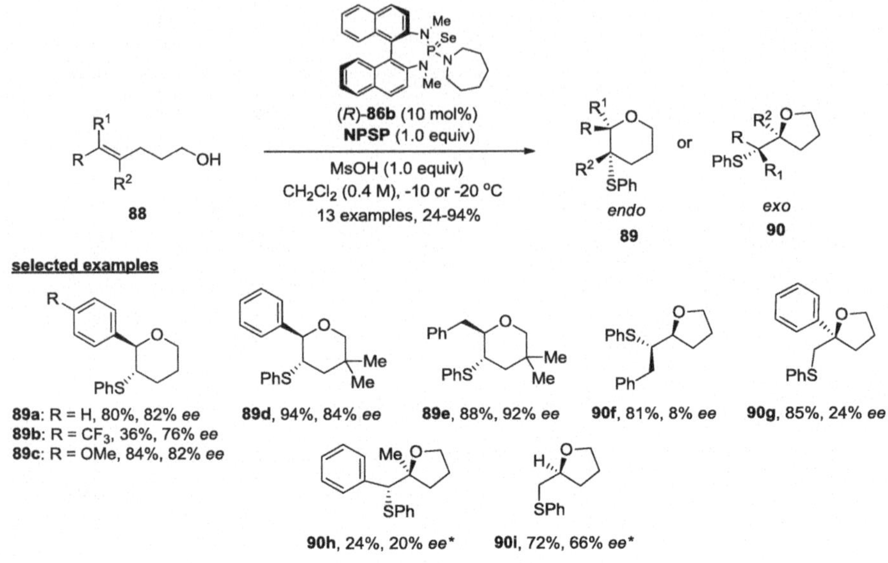

* reaction performed at r.t.

Scheme 1.52 Asymmetric Se-catalyzed thiofunctionalization of unactivated alkenes

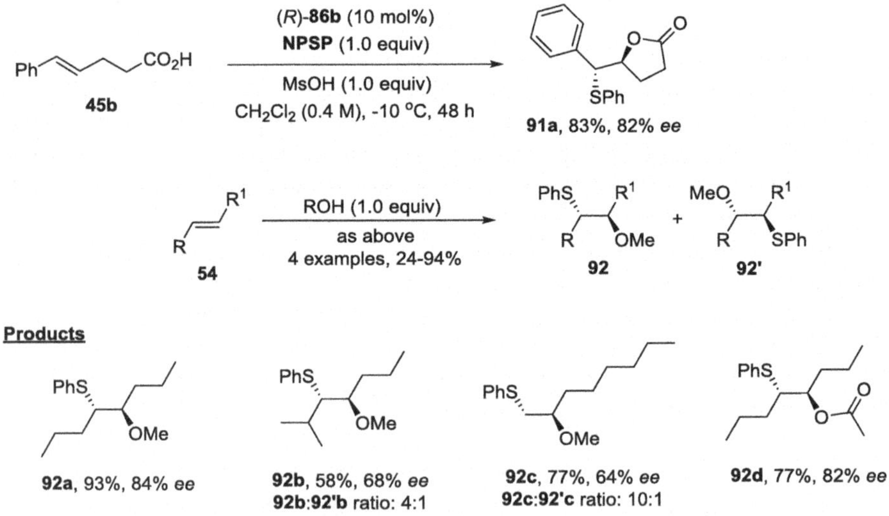

Scheme 1.53 Se-catalyzed thiofunctionalization using internal and exogenous nucleophiles

to deliver the intermediate **C**, that eliminates MsOH, the desired enantioenriched thioethers **89–92** regenerating the catalyst (*R*)-**86b** for a new cycle [85].

The success of BINAM-based selenophosphoramide (*R*)-**86b** as a chiral Lewis base in the activation of NPSP to prepare enantioenriched *S*-THP **89** and *S*-THF **90** from alkenoic alcohols [85], has inspired Denmark to try it in the carbosulfenylation

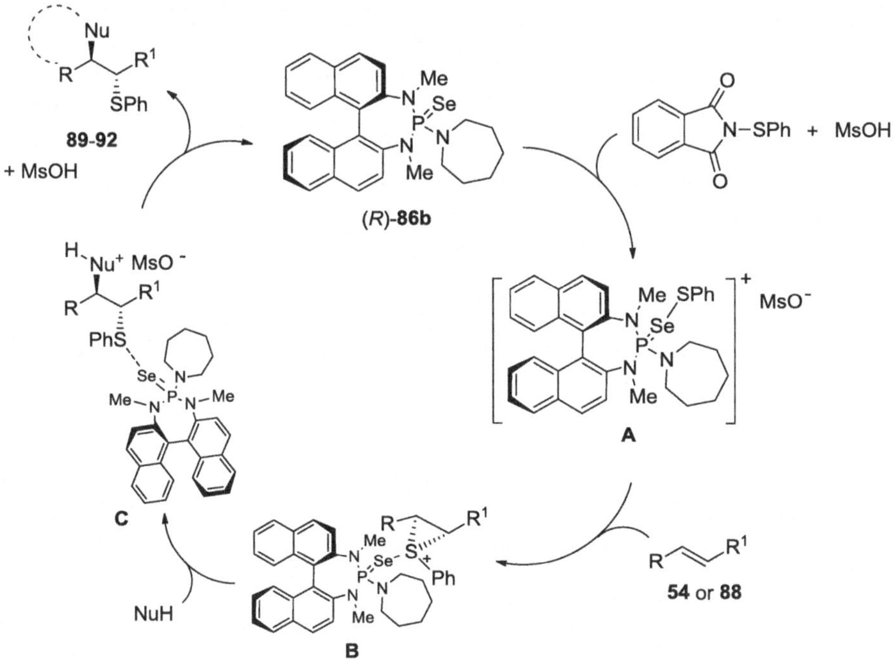

Scheme 1.54 Proposed mechanism for the Se-catalyzed enantioselective thiofunctionalization of alkenes

of alkenes with an aromatic nucleophile [28, 86]. The low activity of (*R*)-**86b** with *cis*-alkenes, however, has motivated the authors to design **84c**, the third generation Denmark's catalyst (Fig. 1.4), a quite similar catalyst, by changing azepane ligand for diisopropyl amine. Catalyst (*S*)-**86c** was successfully used in the asymmetric thiocyclization of alkenes with different pendent nucleophiles (amine, aniline, phenol), affording a range of enantioenriched *N*- and *O*-heterocycles [87–89].

When alkenes containing a benzo[*d*][1,3]dioxole moiety **93** were subjected to the reaction with NPSP (1.0 equiv.) in the presence of $EtSO_3H$ (0.75 equiv.) and Denmark's catalyst **86c** (10 mol%), the respective *trans*-tetrahydronaphthalenes **94** were obtained [28, 86]. To find the best reaction conditions, the authors have screened different thiophosphoramides and selenophosphoroamides, as well as several Brønsted acids. Despite catalyst (*R*)-**86b** and MsOH presented satisfactory outcomes in reactions using *trans*-olefins, (*S*)-**86c**/$EtSO_3H$ were more efficient when *cis*-alkenes were used and were chosen as the best catalyst/acid pair for this carbosulfenylation. In some cases, the better enantioselectivity was obtained at low temperatures (0 or −20 °C), despite the reaction of electron-deficient alkenes were very low and were conducted at room temperature. A total of 17 thio-tetrahydronaphthalenes **94** were obtained in 50–92% yields and up 94% *ee* after 1–6 days of reaction (Scheme 1.55).

The authors extended the reaction to (*E*)-alkenes **93j–l**, with less nucleophilic aryl groups, bearing only one methoxy substituent in the 3- or 4-position respect to

Fig. 1.4 Denmark's catalysts

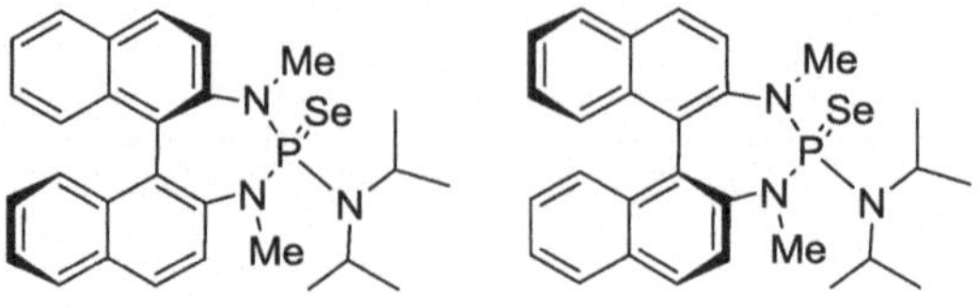

(*R*)-**86c** (*S*)-**86c**

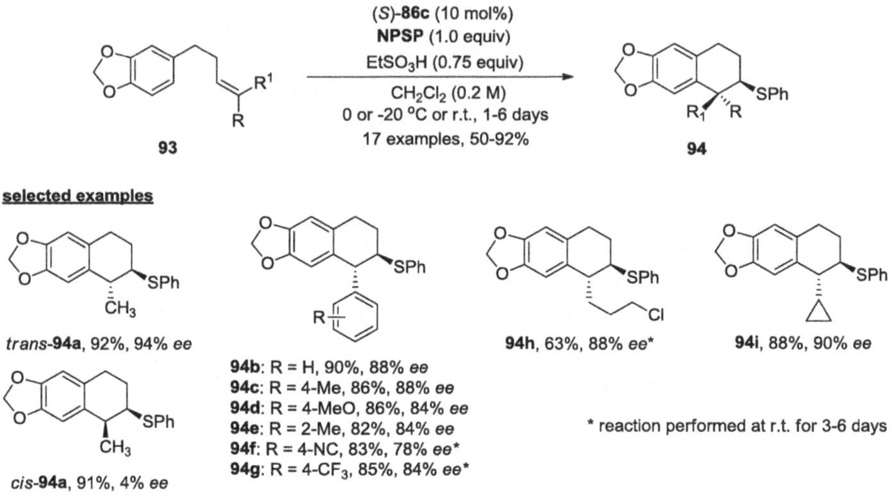

Scheme 1.55 Carbosulfenylation of benzo[*d*][1,3]dioxoles catalyzed by Denmark's catalyst

the tethered alkene (Scheme 1.56). A mixture of cyclized products **95a–b** and **96a–b** was obtained starting from **93j** (R = $^{n}C_5H_{11}$) and **93k** (R = Ph) in 80% and 90% overall yield, respectively and with good enantioselectivity for both isomers of each pair (up 92% *ee*). The only possible cyclization product from *p*-methoxy-substituted aryl **93l** was obtained in 85% yield after 1.5 day at 0 °C, although with a moderate enantioselectivity (68% *ee*). The proposed mechanism for the carbosulfenylation is essentially the same depicted in Scheme 1.53, with the formation of a chiral thiiranium intermediate [28]. For those who are interested in more details on the mechanism of this and other Lewis base-catalyzed asymmetric thiocyclizations developed using catalyst **86** and NPSP, two papers from Denmark are noteworthy, both published in 2014 [90, 91].

Alkenyl sulfonamides **98** (RNH-Ts) were efficiently cyclized in the presence of NPSP and Denmark's catalyst (*S*)-**86c** and MsOH as a coactivator [88]. By the reaction of terminal and internal alkenes **98**, enantioenriched *anti*-2,3-disubstituted piperidines **99**, pyrrolidines **100**, and azepanes **101** were prepared in 39–93% yields and up 95% *ee*. The authors observed that the amount of MsOH (0.5 equiv.) was crucial to avoid the isomerization of piperidine **99** (the kinetic product) to pyrrolidine **100** (the thermodynamic product), that was observed when 1.0 equiv. was

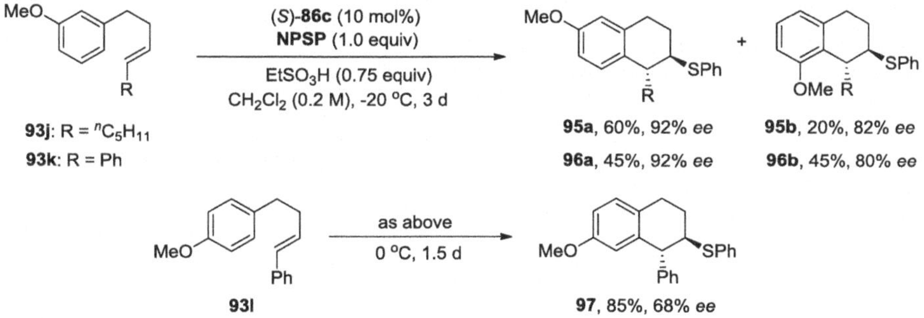

Scheme 1.56 Carbosulfenylation of methoxy arenes catalyzed by Denmark's catalyst

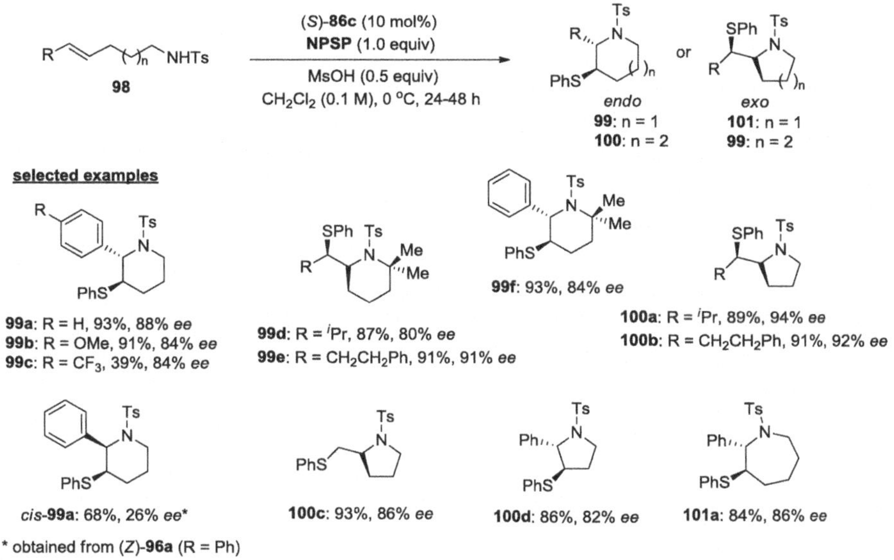

Scheme 1.57 Cyclization of Alkenyl sulfonamides

used (Scheme 1.57). Good yields and high enantioselectivity were obtained for most of the substrates in reactions conducted at 0 °C for 24–48 h. One exception was olefin **98c**, bearing a CF_3 in the aromatic ring, that afforded the respective piperidine **99c** in only 39% yield after 48 h. Mechanistically, this sulfenoamination follows the same catalytic cycle described for the carbosulfenylation showed in Scheme 1.54.

By substituting amines **98** by anilines **102** as the nucleophile, Denmark and Chi were able to prepare *anti*-2,3-disubstituted tetrahydroquinolines **103**, indolines **104**, and tetrahydrobenzazepines **105**, via the Se-catalyzed sulfenoamination using Denmark's catalyst **86c** [89]. The *N*-heterocycle type is a consequence of the length of the tethered double bond from the aromatic ring of aniline **102**. The reaction conditions are essentially the same used in the cyclization of aliphatic tosylamines **98**, except that the overall concentration (in CH_2Cl_2) was increased do 0.4 M (Scheme 1.58).

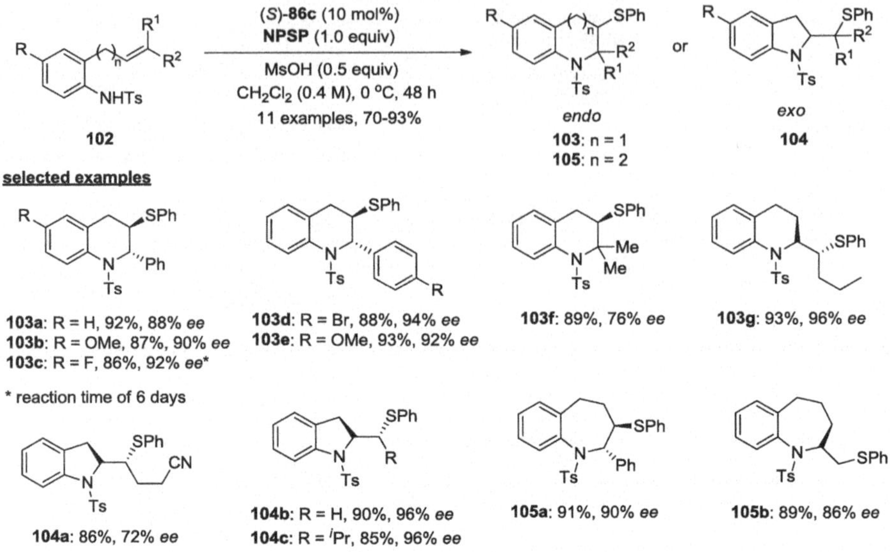

Scheme 1.58 Se-catalyzed sulfenoamination reaction

The authors observed good yields (70–93%) and excellent selectivity (up 96% *ee*) for all substrates after 48 h of reaction at 0 °C, except for aniline **102c** (R = F, n = 1), which required 6 days of reaction to afford the respective *anti*-2,3-disubstituted tetrahydroquinolines **103c** in 86% yield and 92% *ee*. The presence of the electron-donor group OMe in **102b**, however, did not accelerate the reaction compared to neutral one (H in **102a**). This behavior was rationalized by the authors in terms of a change in the turnover-limiting step (TOLS) of the reaction. For electron-rich anilines the formation of chiral thiiranium ion **B** is the TOLS, while for 4-fluoroaniline derivative **102c**, the nucleophilic capture of **B** to form the protonated product is the slow step of the reaction (Scheme 1.59). The authors observed that when an amine group is present in the pendant alkene, it competes for the chiral thiiranium ion **B**, reacting faster than the aniline.

Denmark catalyst (*S*)-**86c** was successfully used by himself and Kornfilt in the intermolecular sulfenofunctionalization of alkenes with phenols to prepare 2,3-disubstituted benzopyrans **107**, dihydrobenzofurans **108**, and benzoxepines **109** in good yields and high enantioselectivity [87]. When a mixture of (*E*)-2-cinnamylphenols **106**, NPSP (1.0 equiv.), and MsOH (0.25 equiv.) in CH_2Cl_2 were reacted in the presence of **86c** (10 mol%), *endo* products benzopyrans **107** were obtained in 70–89% yields. Reactions were conducted for 24–36 h at −20 °C. Electron-rich phenols presented a similar reactivity compared to unsubstituted ones, while the presence of electron-releasing group in the phenol ring decreased the reactivity, as observed for aniline **102c**. The highly electron-deficient *p*-CF_3-substituted cinnamylphenol **106i** presented low reactivity under the optimal conditions using NPSP as sulfur source. Good results, however, were achieved when the reaction was conducted at room temperature for 12 h and *N*-2,6-diisoprop

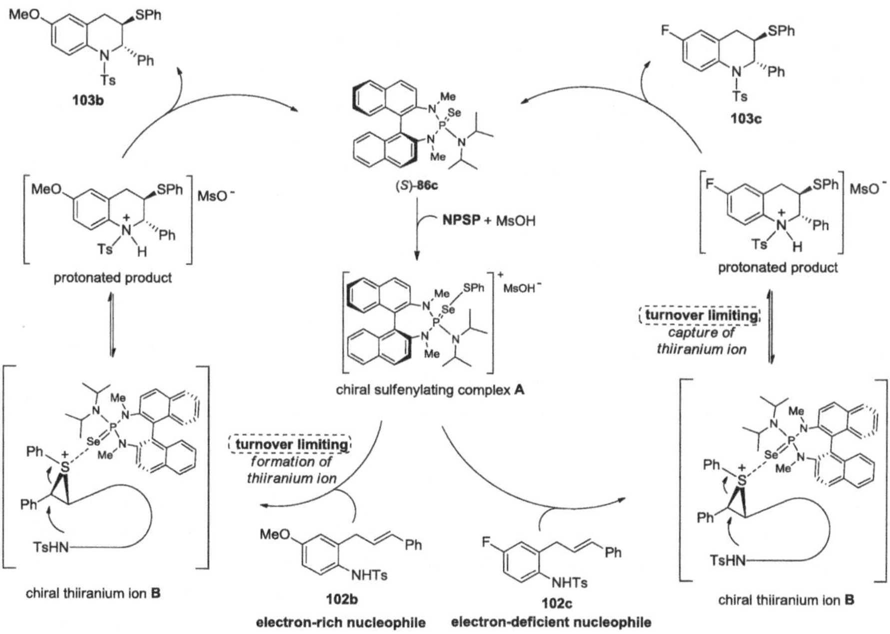

Scheme 1.59 Mechanism of the Se-catalyzed sulfenoamination

ylphenylthiophthalimide (NPrPSP) was used as sulfur source, with **107j** being obtained in 89% yield and 90% *ee* (Scheme 1.60).

The sulfenofunctionalization reaction was efficiently proved using phenols with different substituents in the tethered double bond, like heteroaryl, alkyl, ester and ether groups. The substitution pattern of the double bond and the distance between the double bond and the phenol ring, directly influenced the reactivity of alkene **106** and the identity of the *O*-heterocyclic. The lack of reactivity and low selectivity observed in some substrates using the standard conditions of Scheme 1.60 could be overcome using higher temperature (0 °C) combined with more Brønsted acid (0.5 equiv.) and the use of NPrPSP as sulfenylating agent. The sterically bulky 2,6-diisopropylphenylthio group prevents the erosion of the chiral thiiranium intermediate at the relatively high temperature used with less reactive phenols (Scheme 1.61). Interestingly, if phenol **106** possess a pendant alcohol or carboxylic acid group in the tethered alkene, these react faster than the phenol with the thiiranium intermediate, and THF, THP, or butyrolactone are preferentially obtained.

Catalyst (*R*)-**86c** was successfully used by Denmark and coworkers in the first catalytic, enantioselective sulfenylation of silyl enol ethers. The products, (*S*)-α-phenylthio ketones, were obtained in good yields and excellent enantioselectivity [92]. The reaction conditions were determined after a detailed screening of all the parameters involved in the reaction, such as temperature and the need or not of a Brønsted acid to activate the sulfenylating agent. Because the high sensitivity of silyl enol ethers against acidic hydrolysis, several weak acids were tested, but a

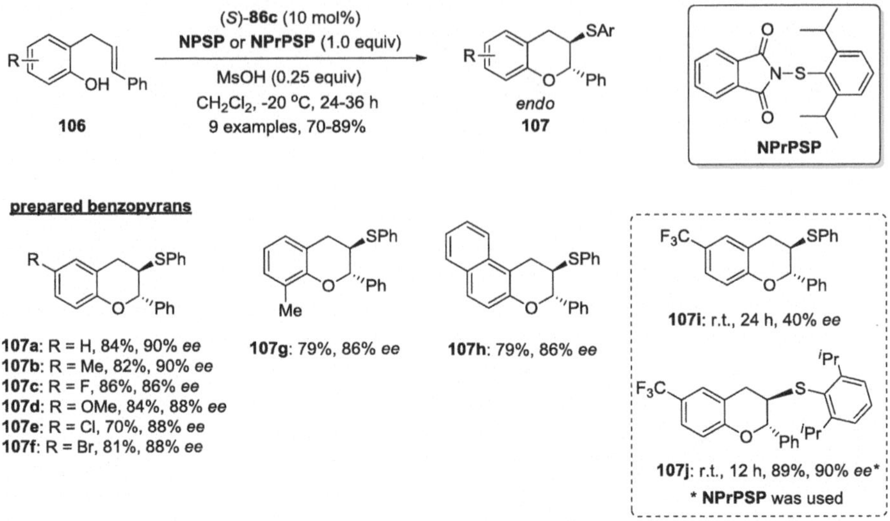

Scheme 1.60 Sulfenofunctionalization of alkenes with phenols

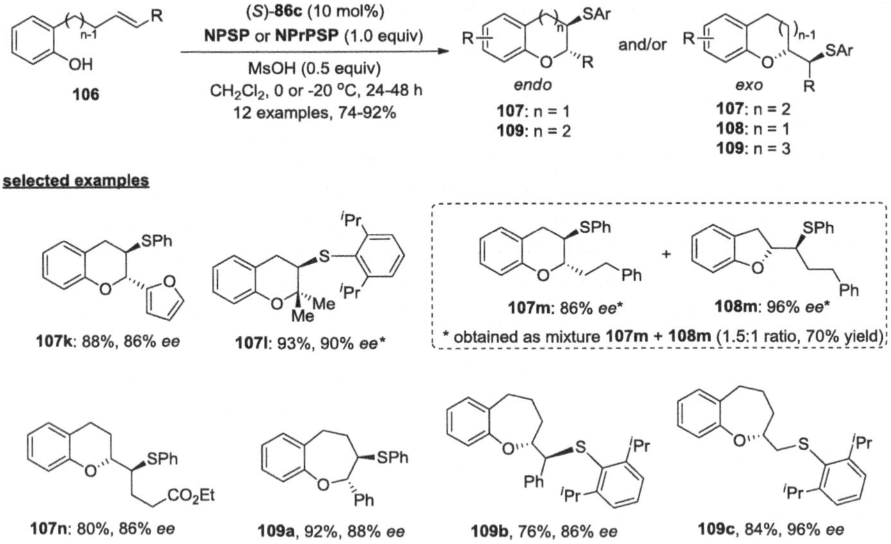

Scheme 1.61 Sulfenofunctionalization reaction catalyzed by **86c**

substantial amount of background reaction to afford racemic thioketone was observed even at low temperatures. The use of *N*-phenylthiosaccharin (NPSSac) as a sulfenylating agent allowed the sulfenylation at neutral conditions at −78 °C, consequently affording high level of enantioselectivity when (*R*)-**86c** (15 mol%) was present (Scheme 1.62). To prevent a racemic background reaction during the workup, the authors developed a quench procedure involving the addition of a pre-

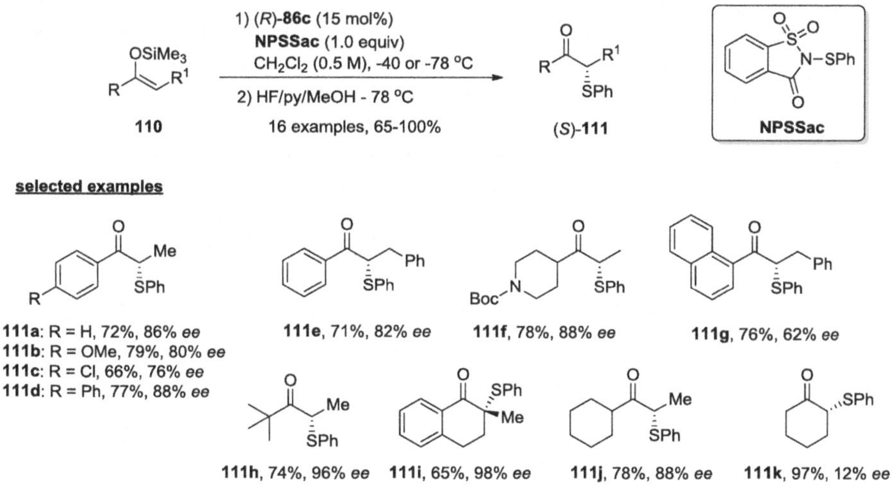

Scheme 1.62 Catalytic, enantioselective sulfenylation of silyl enol ethers

cooled (−78 °C) solution of HF/pyridine in methanol into the reaction mixture at −78 °C. Using this procedure, the maximum enantioselectivity was reached (up to 96% *ee*). The reaction was enantioselective only starting from (*Z*)-silyl enol ethers. This outcome is in agreement with that observed in the carbosulfenylation reaction using the same catalyst **86c** [28, 86] which worked well only using (*E*)-olefins. A close look in the (*Z*)-silyl enol ether structure shows that the carbon-based substituents on the double bond are effectively *trans*.

The chiral selenide **113a** was designed by Yeung and coworkers and used as a Lewis base catalyst in an unprecedented enatioselective bromoaminocyclization of trisubstituted benzenosulfonamides **112** to afford enantioenriched pyrrolidine derivatives **114** [93]. Several chiral dialkyl selenides **113** were prepared in five steps from mannitol and tested for their catalytic Lewis base activation of brominating agents in the reaction with **112a**, aiming to obtain the best yields and higher *ee*. The best result was obtained when 20 mol% of selenosugar **113a** and 1.1 equiv. of *N*-bromophthalimide (NBP) were used in a 1:1 mixture of CH_2Cl_2 and $PhCH_3$ as solvent, with the expected pyrrolidine **114a** being obtained in 90% yield and 92% *ee* after 5 days of reaction at −78 °C (Scheme 1.63). The method was successfully extended to 18 differently substituted sulfonamides **112**, affording in most of cases the respective pyrrolidines in good yields and excellent enantioselectivity, except when strongly electron-rich and electron-poor aryl R^1 groups where attached to the double bond (e.g., 4-ClPh, 4-CF_3Ph and 4-MeOPh).

The first step in the reaction is the interaction of selenide **113a** with NBP, giving the activated brominating species **A** (Scheme 1.64) The interaction of **A** with the double bond of sulfonamide **112** forms the selenium-coordinate bromonium cation **B**, that undergoes an intramolecular attack by the nitrogen of sulfonamide to afford product **114** and phthalimide. The authors proposed that the tight pair Br^+---SeR_2* can minimize the bromine degeneration and racemization, explaining the high enan-

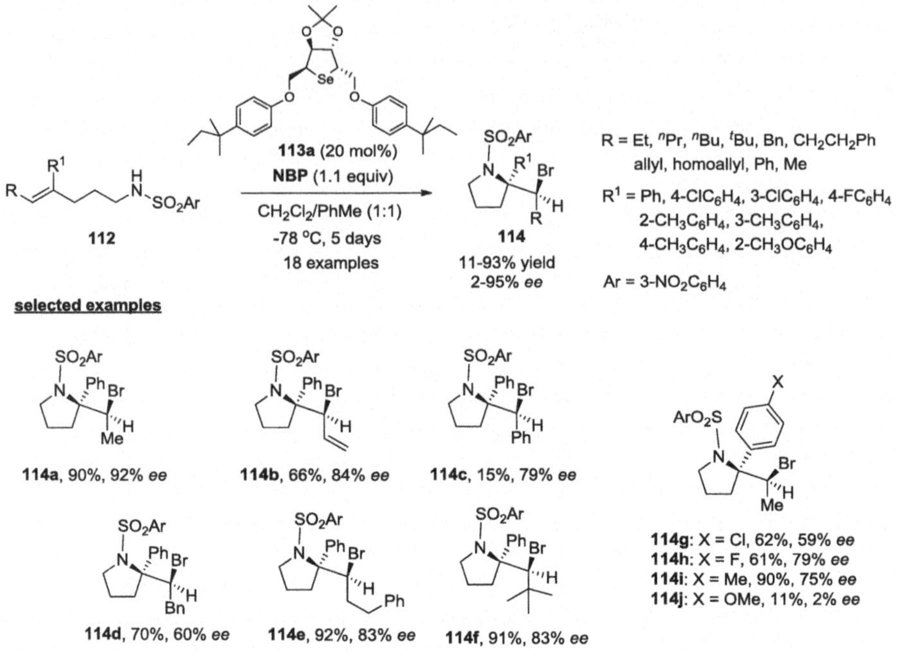

Scheme 1.63 Enantioselective bromoaminocyclization of trisubstituted benzenosulfonamides

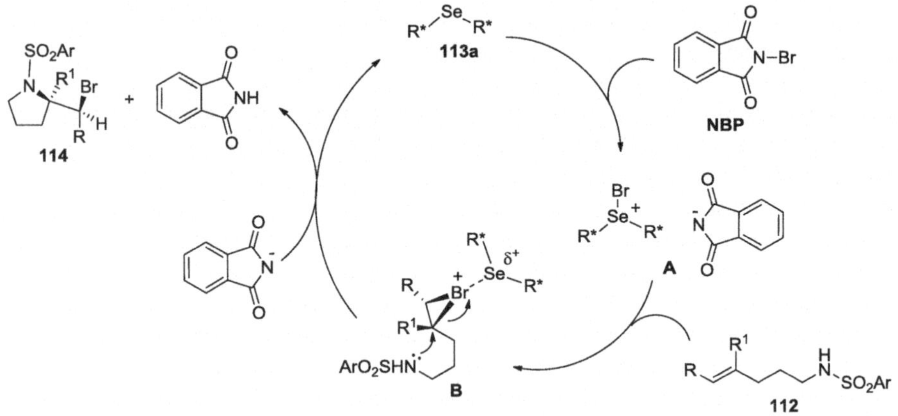

Scheme 1.64 Mechanism of the enantioselective bromoaminocyclization of trisubstituted benzenosulfonamides

tioselectivity. The regioselectivity (Markovnikov) and the stereoselectivity *anti* are a consequence of the dominant role of the bromonium intermediate **B**.

Zhao and coworkers used a similar strategy to prepare enantioenriched trifluoromethylthiolated pyrrolidines **117** and piperidine **118** starting from unsaturated sulfonamides **112** and **115** [94]. The authors have prepared and tested 13 chiral

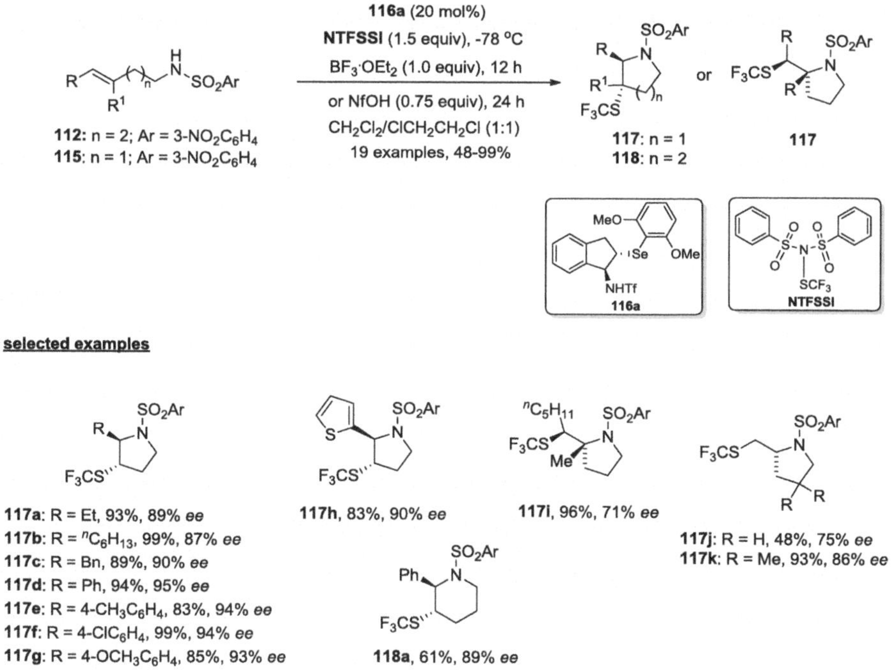

Scheme 1.65 Se-catalyzed preparation of enantioenriched trifluoromethylthiolated pyrrolidines and piperidine

indane-based bifunctional selenides **116** and compared them with themselves and with the sulfur analogues in the cyclization of sulfonamide **112k** ($n = 2$, R = Et, R^1 = H) with *N*-(trifluoromethyl)-thiobenzenesulfonamide (NTFSSI, 1.5 equiv.). The best results were obtained using 20 mol% of the catalyst **116a** in the presence of TfOH (0.5 equiv.) in CH_2Cl_2 at −78 °C, with the desired trifluoromethylthiolated pyrrolidine **117a** (R = Et) being obtained in 97% yield and 81% *ee* after 12 h. In a general way, the chiral selenides delivered higher yields and better selectivity of **117a** than the sulfur analogues. After a screening for the best acid and solvent for the reaction, it was observed that $BF_3{\cdot}Et_2O$ (1 equiv.) in a 1:1 mixture of CH_2Cl_2 and $ClCH_2CH_2Cl$ afforded **117a** in 93% yield and 89% *ee* (Scheme 1.65). The optimal conditions were used to prepare 19 differently substituted trifluoromethylthiolated pyrrolidines **118** in 48–99% yields and 71–97% *ee*. In some cases, 0.5 equiv. of the Brønsted acid 1,1,2,2,3,3,4,4,4-nonafluoro-1-butanesulfonic acid (NfOH) was used instead $BF_3{\cdot}Et_2O$. When (*E*)-5-phenylpent-4-en-sulfonamide **112l** (R = Ph, R^1 = H) was used as starting material, six-membered trifluoromethylthiolated piperidine **118a** was obtained in 61% yield and 89% *ee* after reaction with NTFSSI in the presence of NfOH as Brønsted acid. Despite a mechanism was not depicted in the paper, reaction conditions indicate that it most likely is close related to that showed in Scheme 1.65, for the bromoamination reaction.

1.5 Organoselenium Compounds as Activator of Hydrogen Peroxide

1.5.1 Hydrogen Peroxide in Oxygen-Transfer Reactions

Currently, a number of papers and book chapters have been dedicated to the environmental advantages of using hydrogen peroxide as an atom-efficient, green oxidant agent [95]. Differently from most oxidant used in organic synthesis, no large amounts of salt are generated after the reaction with H_2O_2 and the only side product in the reaction is nonhazardous water, that most of times can be easily separated from the product by decantation. A drawback in using hydrogen peroxide is the requirement of a promoter or catalyst to allow the oxygen-transfer, once many of the reactions involving H_2O_2 are limited by the kinetics of the reaction. To circumvent the low reactivity of H_2O_2, transition metal-based [96–100], organic [101], and inorganic [102] catalysts have been used.

One of the remarkable characteristic of organoselenium compounds is their ability to be oxidized by H_2O_2 and organic hydroperoxides at very mild conditions [11–17]. The formation of selenoxides by the reaction of diorganyl selenides with H_2O_2 is the key step in the classical *syn*-elimination of selenoxides, one of the milder way to access alkenes, discovered by Sharpless in 1973 [103]. When diaryl diselenides are reacted with H_2O_2, the identity of the product depends on the amount of H_2O_2 used in the reaction. Selenenic, seleninic, or perseleninic acids and selenenic anhydride can be formed from diselenide, and normally it is accepted that these are the active species that transfer an oxygen atom to the substrate (Fig. 1.5) [12–14].

The H_2O_2 activation by diselenides and ebselen derivatives has been used in a range of oxygen-transfer reactions, including the oxidation of carbonyl, azomethine, and organosulfur compounds, benzilamines, arenes, and alkenes. Epoxidation and Baeyer–Villiger reactions, dehydration of aldoximes to prepare nitriles and the generation in situ of bromine or electrophilic Br species are other examples of the plethora of preparative methods using organoselenium as catalysts [11–13, 15, 16].

1.5.1.1 Oxidation of Aldehydes to Carboxylic Acids and Esters

Se-catalyzed oxygen-transfer reactions using hydrogen peroxide were made even more green by Santi and coworkers, who described a Se-catalyzed green protocol to prepare carboxylic acids **120** and esters **121** starting from aldehydes **119** using aqueous hydrogen peroxide (1 equiv.) as an oxidant and $(PhSe)_2$ **1a** (2 mol%) as the catalyst [104]. The identity of the product (acid or ester) was determined by the solvent (water or alcohol) used in the reaction. When water was the solvent, 12 different aryl-, alkenyl-, and alkyl-carboxylic acids **120** were obtained in 75 to >99% yields after 3–6 h of reaction at room temperature. The only exception was *p*-nitrobenzaldehyde **119e**, that required 24 h to afford the respective *p*-nitrobenzoic acid **120e** in 88% yield (30% in 6 h) (Scheme 1.66). Excellent yields (>99%) of

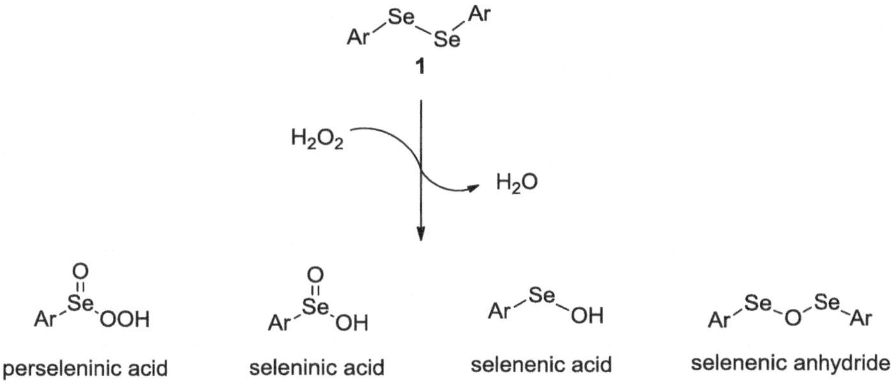

Fig. 1.5 Products of the oxidation of diselenide by H_2O_2

methyl esters derived from benzaldehyde **119a** and electron-poor *p*-nitrobenzaldehyde **119e** and *m*-trifluoromethylbenzaldehyde **119f** were obtained after 5 h of reaction at 75 °C in methanol. Unfortunately, electron-rich *p*-tolualdehyde **119c** and *p*-anisaldehyde **119d** did not give satisfactory selectivity, delivering the expected methyl esters **121c** and **121d** respectively as a mixture with *p*-cresol and *p*-methoxyphenol via a competitive Dakin-like reaction. Benzaldehyde was efficiently reacted with other alcohols, including branched and cyclic ones, with good results (Scheme 1.66). The greenness of the method was increased by performing the reaction of benzaldehyde in gram scale; benzoic acid thus obtained was filtered off, and the mixture of solvent (water) and the catalyst **122a** was reused in additional four reactions without significant loss in the activity.

In the mechanism proposed by the authors, benzeneseleninic acid **123** and benzeneperseleninic acid **122** are the actual species participating in the catalytic cycle of the reaction, with the oxygen transfer to the aldehyde **119** occurring from **123a**. The involvement of **123a**, generated in situ in the reaction of **1a** with H_2O_2, was confirmed when the reaction was repeated using 4 mol% of preformed **123a**, giving around the same yields both in the formation of carboxylic acids and esters (Scheme 1.67).

In some cases, it is possible to improve the recycling efficiency and extend the life of the organoselenium catalyst if it is supported on a solid. Polystyrene-supported phenylseleninic acid was efficiently used by Taylor and Flood as a catalyst in the synthesis of 1,2-diols from alkenes and in the Baeyer–Villiger reaction of cyclic and alicyclic alkenes using H_2O_2 [105]. Młochowski and coworkers prepared a silica-supported benzisoselenazol-3(2*H*)-one analogue and used it in the catalytic activation of H_2O_2 and ^{t}BuOOH (TBHP) to prepare sulfoxides and sulfones from diorganyl sulfides [106]. A different approach for the oxidation of aldehydes in aqueous media was developed by Nemati and coworkers, using the heterogeneous catalyst **124**, with the diaryl diselenide moiety incorporated on the surface of SiO_2-coated Fe_3O_4 magnetic nanoparticles [107]. The Se-containing catalyst **124** was

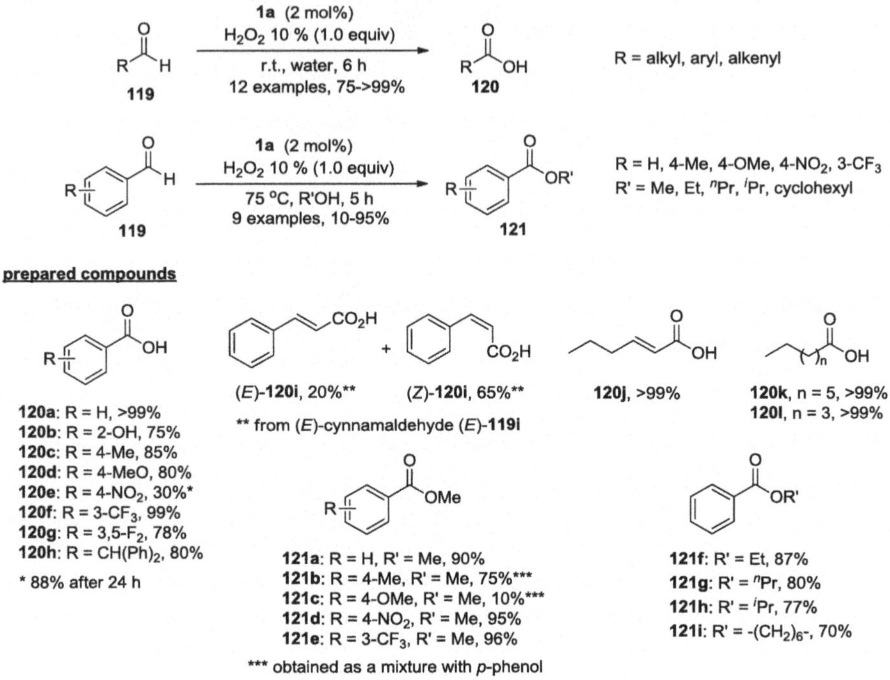

Scheme 1.66 Se-catalyzed synthesis of carboxylic acids and esters

Scheme 1.67 Conversion of the precatalyst to the actual catalytic species

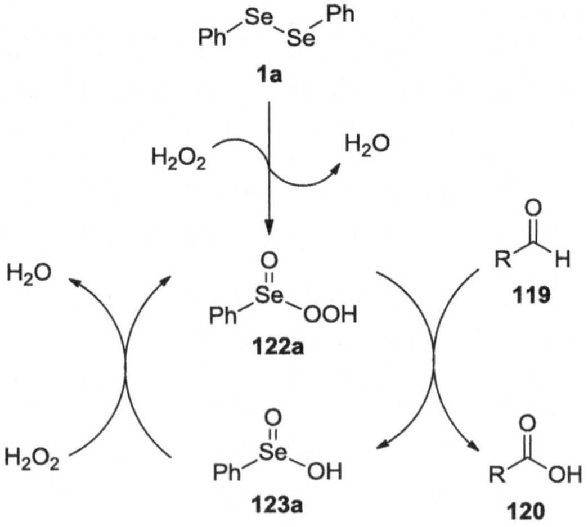

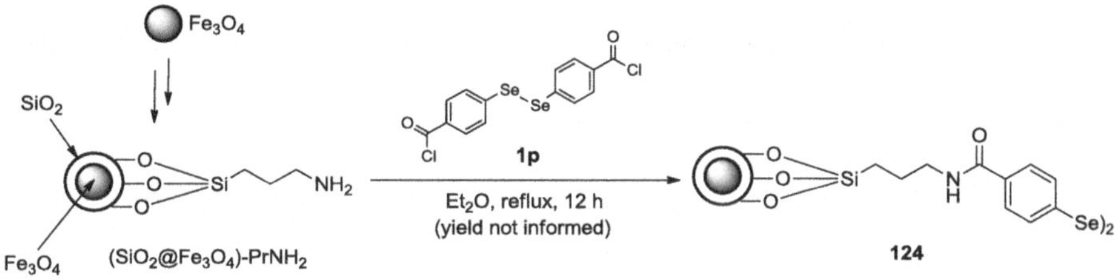

Scheme 1.68 Preparation of the catalyst **124**

prepared by the reaction of the aminopropyl derivative of silica encapsulated magnetite nanoparticles (SiO_2@Fe_3O_4)-PrNH_2 with 4,4′-diselanediyldibenzoyl chloride **1p** (Scheme 1.68).

Catalyst **124** (loading of 20 mg/mmol of substrate) was successfully used in the oxidation of aromatic and aliphatic aldehydes **119** to the respective carboxylic acids **120** in yields from 75 to 98% in the presence of H_2O_2 30% (4.9 equiv.) at 60 °C without any solvent (Scheme 1.69). Similar to the observed by Santi and coworkers in the aqueous reaction (Scheme 1.68), the worst yields were obtained starting from electron-deficient benzaldehyde derivatives (50–75%). In contrast to the aforementioned work, (*E*)-cinnamaldehyde **119i** afforded exclusively (*E*)-cinnamic acid **120i**, without isomerization product (Scheme 1.69).

The mechanism proposed by the SiO_2@Fe_3O_4-anchored diselenide **124** is essentially the same described by Santi and others (Scheme 1.67) and involves the oxygen-transfer to the aldehyde by the respective anchored perseleninic acid. In a catalyst reuse test, after completion of the reaction of benzaldehyde **119a** with H_2O_2 the catalyst was easily removed from the reaction mixture using a permanent external magnet. After washing with ethanol and acetone, the catalyst was dried and could be reused up four times with minimal decrease in the activity in the fourth reaction.

1.5.1.2 Synthesis of Ketones by C=C Bond Cleavage

In 2017, a hitherto unprecedented Se-catalyzed oxidative C=C bond cleavage by an excess of H_2O_2 was developed by Yu and coworkers to prepare ketones [108]. Among several diaryl and dialkyl diselenides tested as the catalyst for the reaction, the best results were obtained using 5 mol% of dicyclohexyl diselenide **1q** (or, in some cases, dibenzyl diselenide **1o** or dibutyl diselenide **1r**) for 48–96 h in ethanol under reflux using H_2O_2 30% (5 equiv.). The respective symmetrical and unsymmetrical ketones **125** were obtained in 30–74% yields starting from *gem*-disubstituted and 1,1,2-trisubstituted alkenes **54** (Scheme 1.70). The authors have observed that the reaction did not work with tetrasubstituted alkenes and that in some cases, temperatures so high as 120 °C were necessary (a sealed tube was used).

The proposed mechanism for the formation of ketones **125** from disubstituted and trisubstituted alkenes **54** starts by the Se-catalyzed formation of epoxide **A** via radical or electrophilic mechanisms. Epoxide **A** is then hydrolyzed to the

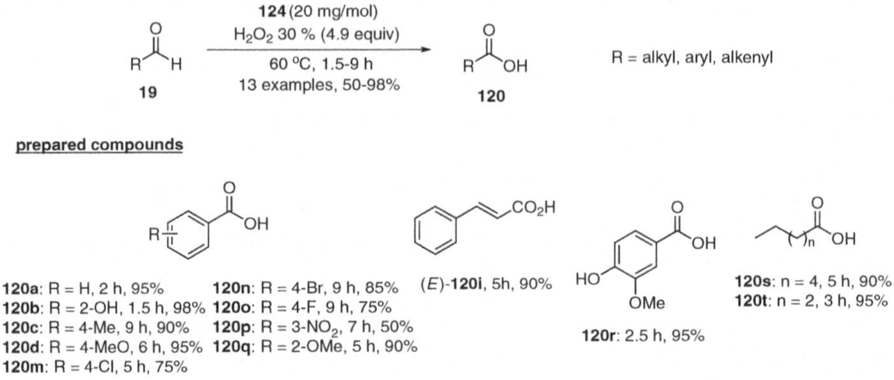

Scheme 1.69 Oxidation of aldehydes using catalyst **124**

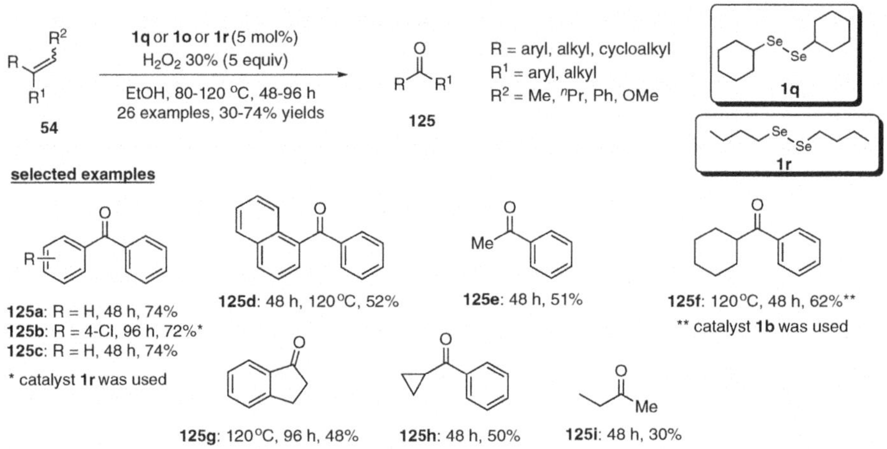

Scheme 1.70 Se-catalyzed oxidative C=C bond cleavage to prepare ketones

1,2-diol **B**, which is oxidized to the carbonyl products by two catalytic cycles. In the cycle **I**, diselenide **1** is oxidized by H_2O_2 to perseleninic acid **C**, the active species, which dehydrates with **B** at the R^2 rather than R, R^1 site hydroxyl, due to less steric hindrance, affording **D**. Rearrangement in intermediate **D** gives hydroxy ketone **E** and seleninic acid **F**, which could be oxidized by H_2O_2 to regenerate perseleninic acid **C** for a new cycle (Scheme 1.71). In cycle **II**, a nucleophilic addition of **C** to intermediate **E** affords the organoselenium peroxide **G**, that after a rearrangement forms intermediate **H** and seleninic acid **F**. Further hydration of **H** delivers the final ketone **125** and carboxylic acids, observed in the reaction as their ethyl esters, from reaction with the solvent ethanol. The difficulty in forming the key intermediate **E** from tetrasubstituted alkenes can explain why those alkenes failed in the reaction.

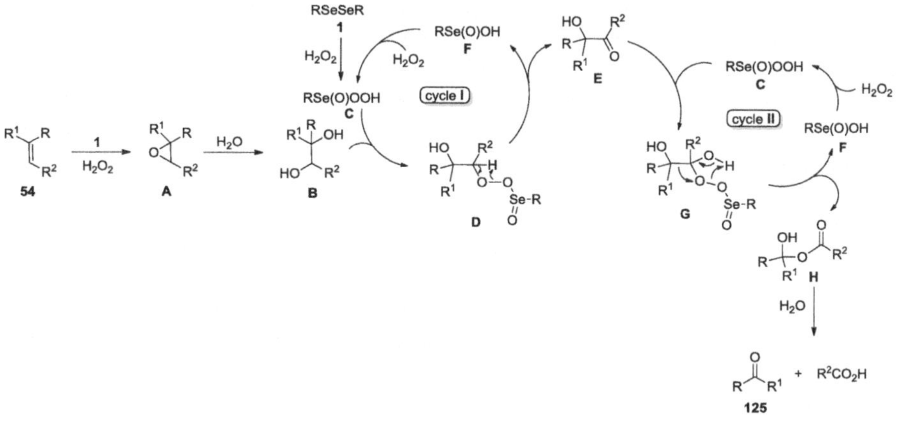

Scheme 1.71 Proposed mechanism for the formation of ketones **125**

1.5.1.3 Baeyer–Villiger and Cognate Reactions

Since the pioneering works from Syper in the Se-catalyzed Baeyer–Villiger reaction (BV) using hydrogen peroxide as oxidant [109], several advances were achieved, especially in terms of the complexity and scope of substrates that have been used in the reaction. Three interesting works published in 2014 and 2015 are good examples of the versatility of Se-catalyzed BV reactions in the synthesis of valuable compounds that are difficult to prepare by conventional methods [44, 110, 111].

In this line, Yu and coworkers developed a new method to prepare (*E*)-2-methylenecyclobutanones (2-MCBones) **126** and used this nontrivial scaffold as a substrate in the Se-catalyzed Baeyer–Villiger oxidation with H_2O_2 to prepare γ-butyrolactones **127** [44]. The authors have tested elemental selenium, selenium dioxide, diphenyl disulfide, and diphenyl diselenide **1a** as possible catalysts for the reaction at different temperatures and catalyst loadings. The best results were obtained using 5 mol% of $(PhSe)_2$ **1a** in the presence of H_2O_2 30% (5 equiv.) at room temperature for 24 h, affording the respective methylene γ-butyrolactones **127** in 40–82% yields (Scheme 1.72).

The authors have performed some test reactions and ^{77}Se-NMR analysis to acquire subsidies to a plausible mechanism for the BV oxidation of 2-MCBones **126** to γ-butyrolactones **127**. They discovered that, differently to what described for other Se-catalyzed BV oxidations using H_2O_2, the first species formed in the oxidation of $(PhSe)_2$ **1a** is benzeneperseleninic anhydride **128a** (detected by ^{77}Se-NMR) instead benzeneseleninic acid **123a**. The hydrolysis of **127a** delivers the active oxidant benzeneperseleninic acid **122a**, which nucleophilically adds to the carbonyl of **126**, affording intermediate **A**. The higher reactivity of the carbonyl compared to the exocyclic double bond blocks the formation of epoxide intermediate **B**. Following a rearrangement of the preferred intermediate **A** in a regiospecific form occurs, to give the desired (*E*)-4-methylenebutenolide **127**. The presence of the conjugated double bond is essential in driving the selectivity of the reaction (Scheme 1.73).

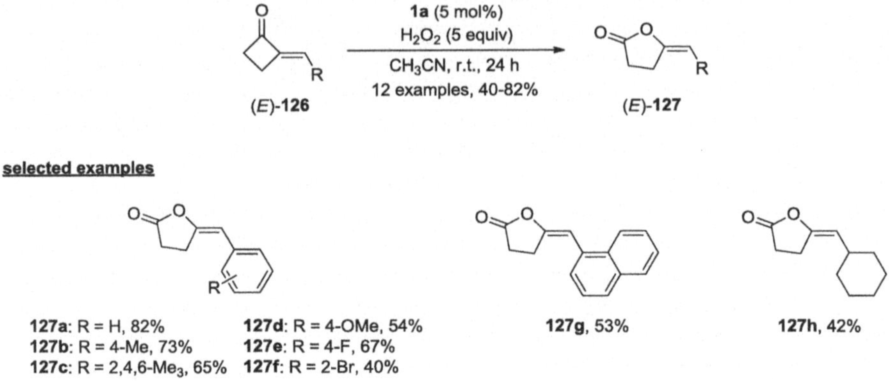

Scheme 1.72 Se-catalyzed Baeyer–Villiger oxidation to prepare γ-butyrolactones

The same research group reported the first Se-catalyzed BV oxidation of heterocycles, involving the preparation of isatoic anhydride **130** from isatin **129** using 5 mol% of $(PhSe)_2$ **1a** as the catalyst and 2 equiv. of H_2O_2 30% as the oxidant [110]. Seventeen isatoic anhydride derivatives with a diversity of substituents at the aromatic ring were selectively prepared in 70–89% yields after 8 h at room temperature in a solution of acetonitrile or a mixture acetonitrile/DMF. The best catalyst (**1a**) was chosen after a screening of 16 different compounds, including diorganyl diselenides, selenides, $(PhS)_2$, $(PhTe)_2$, SeO_2, and PhSeBr. Isatins containing both electron-releasing and electron-withdrawing groups were successfully oxidized under the optimized conditions. For *N*-unsubstituted isatins, best yields were obtained using a 9 vol% DMF/acetonitrile as the solvent (Scheme 1.74).

The mechanism proposed by the authors for the BV reaction of isatin **129** to isatoic anhydride **130** was the same described in Scheme 1.71, with the formation of benzeneperseleninic anhydride **128a** as the active oxidant. The mother liquid of the reaction was reused in new oxidations up seven times with satisfactory yields of **130** (77–94%), simply by adding more **129** and H_2O_2 after removing the product of the previous reaction.

Taking advantage of the Se-catalyzed Baeyer–Villiger oxidation of α,β-unsaturated ketones, Yu and Lu were able to prepare vinyl esters using H_2O_2 as a green oxidant [111]. Starting from differently substituted aromatic or conjugated enones **131**, the respective vinyl esters **132** were obtained using 5 mol% of dibenzyl diselenide **1o** as the catalyst and 4 equiv. of H_2O_2 30% as the oxidant in MeCN. A total of 23 vinyl esters **132** were obtained in 62–88% yields after 24 h at room temperature (Scheme 1.75). Similar to the previous two works of the same group, a possible mechanism could initially involve the oxidation of dibenzyl diselenide **1o** to the respective perseleninic anhydride analogue to **128o**, which after hydrolysis forms the perseleninic acid, the active oxygen transfer in the reaction (see Scheme 1.75).

There are several diselenides that have been designed to be used as catalysts in oxygen-transfer reactions using H_2O_2 as the oxidant to prepare epoxides and Baeyer–Villiger products in a selective and efficient procedure. For example,

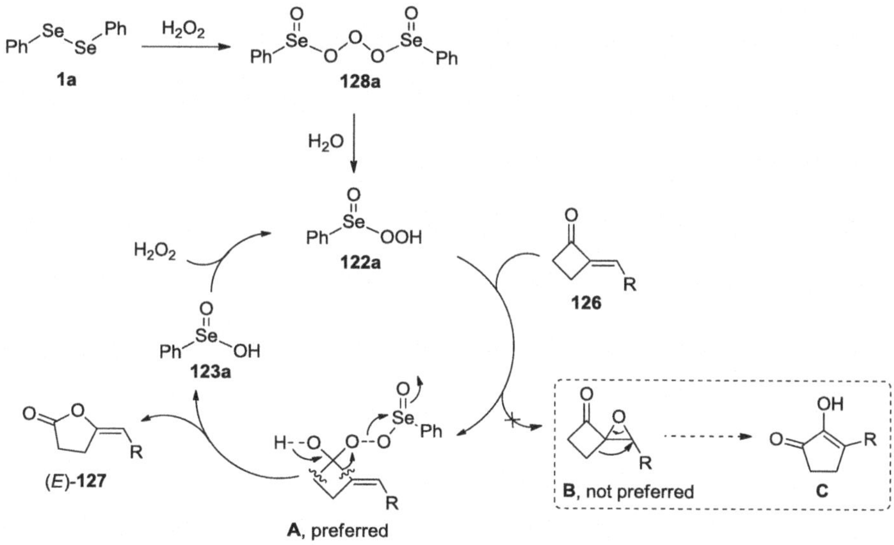

Scheme 1.73 Mechanism for the Se-catalyzed Baeyer–Villiger oxidation

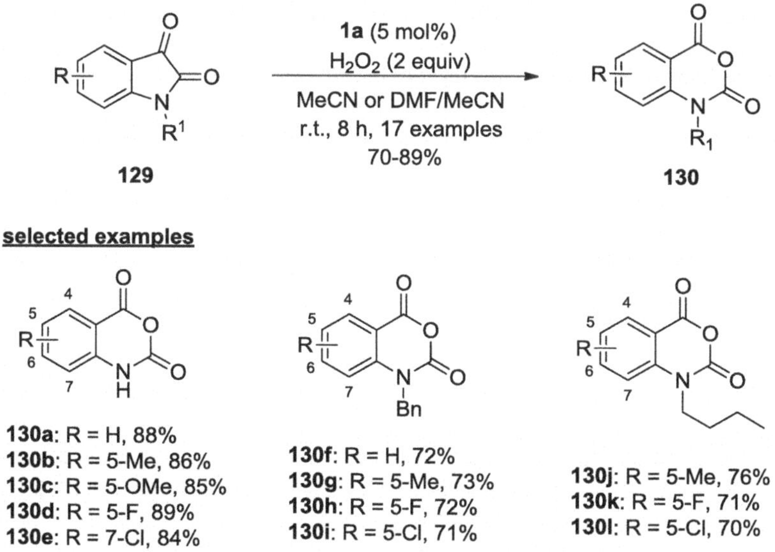

Scheme 1.74 Se-catalyzed preparation of isatoic anhydride **130** from isatin **129**

Sheldon and coworkers used bis[3,5-bis(trifluoromethyl)phenyl] diselenide **1s** as a catalyst in the BV oxidation of cyclic and acyclic ketones **133** and in the oxidation of aldehydes **119** to carboxylic acids **120** [48]. Catalyst **1s** was selected after the analysis of ten diselenides and SeO_2, presenting the best conversion and selectivity compared to the others. The best solvent for the reaction

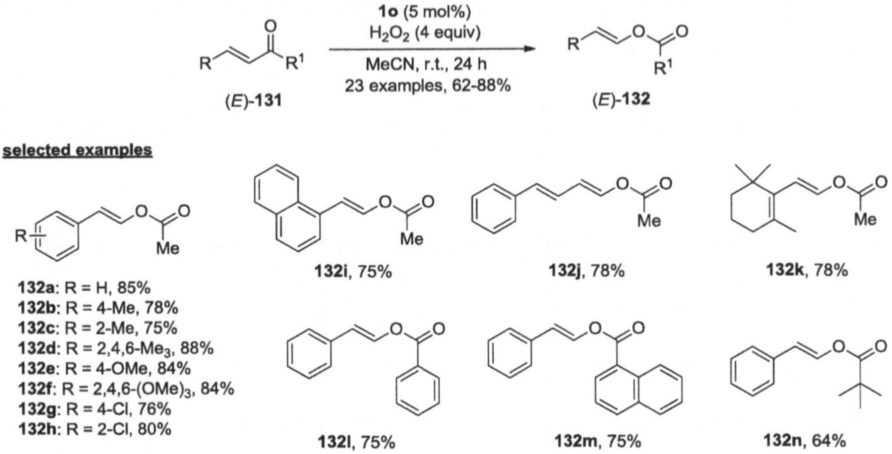

Scheme 1.75 Se-catalyzed Baeyer–Villiger oxidation of α,β-unsaturated ketones

was 1,1,1,3,3,3-hexafluoro2-propanol; however, due its high cost, the authors suggested that the reaction could be conducted in 2,2,2-trifluoroethanol or even CH_2Cl_2. Excellent conversion rates and selectivities were obtained for cyclic ketones; however, *p*-chloroacetophenone and heptan-4-one were less reactive, with only 25% being oxidized after 24 h of reaction to give a 1:1 mixture of the respective BV products and *p*-chlorophenol or butyric acid. Electron-rich 3,4,5-trimethoxybenzaldehyde afforded exclusively 3,4,5-trimethoxyphenol **135a**, together with the ester of formic acid, as a consequence of a fast migration of the phenyl ring. The same reaction was observed for tolualdehyde, but in a less extension (55%) (Scheme 1.76). The mechanism proposed by Sheldon involves at first the formation of a benzeneperseleninic acid **122s** (Schemes 1.67 and 1.73), which could be able to transfer a hydrogen atom to the ketone **133** or the aldehyde **119**.

Ichikawa and coworkers prepared bis[2(trifluoromethanesulfonate)phenyl] diselenide **1t** and used it as a catalyst in the BV oxidation of four different cyclohexanones using H_2O_2 30% as oxidant [47]. The respective caprolactones **134** were obtained in 41–99% yields after stirring a solution of ketone **133** in CH_2Cl_2 in the presence of 5 mol% of catalyst **1t** and 5 equiv. of H_2O_2 30% at room temperature for 16–22 h. The involvement of seleninic acid **122t** as the active species was proved using 2-trifluoromethanesulfonatebenzeneseleninic acid (10 mol%) instead **1t** as the catalyst, affording similar yield of the expected caprolactone **134f** (85% after 16 h) (Scheme 1.77).

1.5.1.4 Synthesis of Epoxides from Alkenes

The epoxidation of olefins by H_2O_2 catalyzed by arylseleninic acids was firstly described in 1977 and 1978 by Reich [112], Grieco [113], Kametani [114], and Sharpless [115]. Catalyst **1s** was used by Sheldon in the epoxidation of a variety

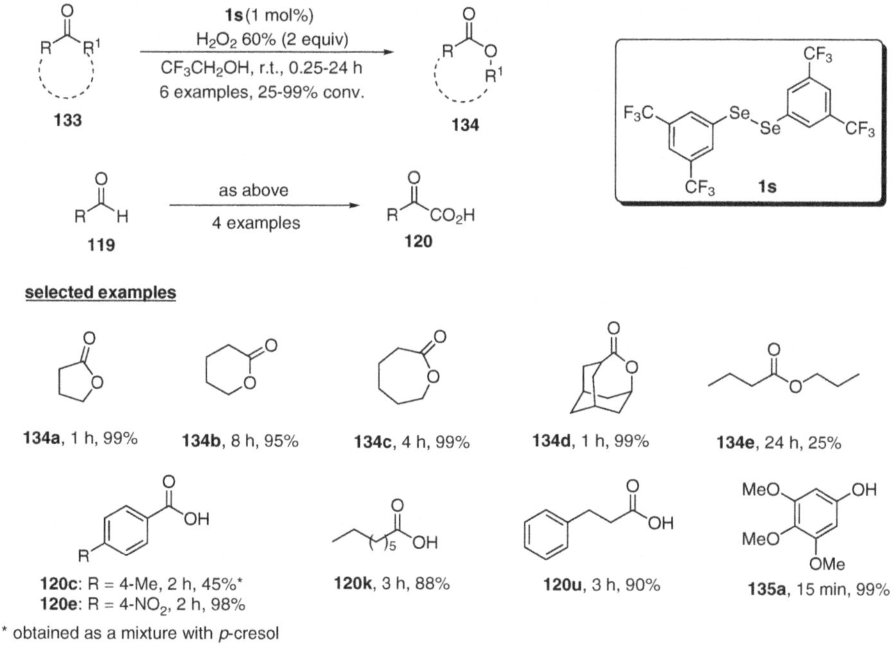

Scheme 1.76 Baeyer–Villiger oxidation of cyclic and acyclic ketones

of isolated internal and terminal alkenes **54** with H_2O_2 60% [116]. The experimental procedure was essentially the same of Scheme 1.74, except that 0.25 mol% of **1s** was used instead 1 mol% and NaOAc (0.2 mol%) was added as a weak base, to neutralize the acidic H_2O_2 solution, minimizing the formation of 1,2-diol, improving the yield of epoxide **136** and reducing the necessary amount of the catalyst (Scheme 1.78).

Yu and coworkers have discovered an interesting, underexplored, peculiarity of Se-catalyzed oxygen-transfer reactions using H_2O_2: by changing the identity of diselenide **1**, they were able to switch the chemoselectivity in the oxidation of β-ionone **137**, driving the reaction to the formation of the epoxide **136i** or the vinyl ester **132o** [117]. It was observed that electron-poor diaryl diselenides, like **1s**, favor BV oxidation, while dibenzyl diselenide **1o** and dibutyl diselenide **1r** afforded preferentially the epoxide **132o**. After a study in the optimization of both reactions, the authors concluded that bis[3,5-bis(trifluoromethyl)phenyl] diselenide **1s** (1 mol%) was the best catalyst for the BV reaction, while dibenzyl diselenide **1o** (5 mol%) deliver the best results in the epoxidation of the deactivated double bond of β-ionone **137** (Scheme 1.79).

The mechanism of the BV reaction and the epoxidation were supposed to involve the same intermediate previously described by the authors in previous works involving Se-catalyzed reactions using H_2O_2 [44, 110, 111]. The first step is the oxidation of the diselenide to perseleninic anhydride **128**, which is hydrolyzed to perseleninic acid **122**, the real catalytic species in the reaction. The mechanism proposed by the

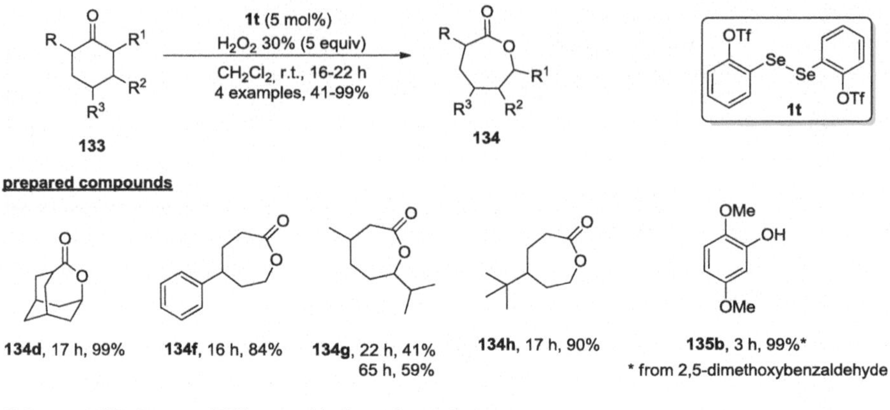

Scheme 1.77 Baeyer–Villiger oxidation of cyclohexanones

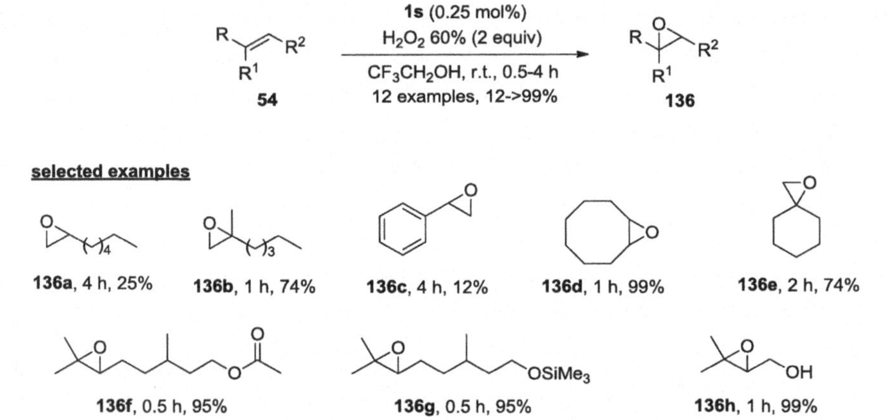

Scheme 1.78 Epoxidation of olefins

Se-catalyzed epoxidation of β-ionone **137** using **1s**/H_2O_2 starts by an electrophilic addition of the electron-deficient Se of perseleninic acid **122s** to the γ,δ-C=C bond of **137**, giving carbocation intermediate **A**, that rapidly forms intermediate **B**. Then, an intramolecular rearrangement of **B** affords epoxide **136i** and seleninic acid **123s**, which is oxidized by H_2O_2 to regenerate perseleninic acid **122s** for a new cycle. The presence of two strong electron-withdrawing CF_3 groups in diselenide **1s** could generate perseleninic acid **122s** with stronger Se^+ electrophilic center than simple diselenides, thus favoring the catalysis for epoxidation (Scheme 1.80A).

In the BV oxidation reaction, the key step involves a nucleophilic attack by one oxygen atom from perseleninic acid **122o**, derived from dibenzyl diselenide **1o**, to the carbonyl of β-ionone **137**, giving intermediate **C**. After an intermolecular rearrangement, **C** releases the vinyl ester **132o** and seleninic acid **123o**, which is reoxidized to **122o** for a new reaction with β-ionone **137**. Again, a close analysis in the mechanism can explain the role of the catalyst structure in the chemoselectivity of

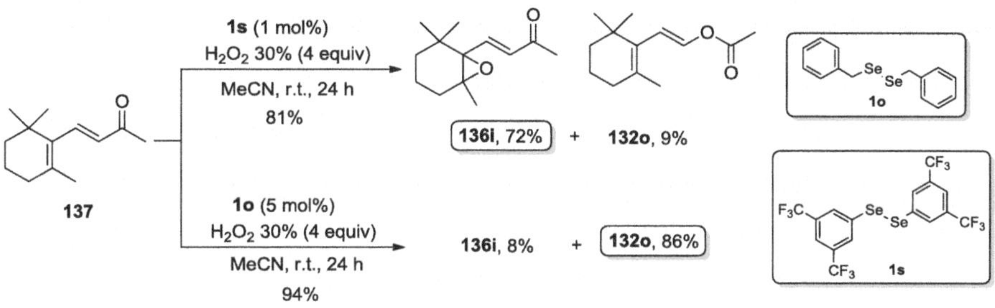

Scheme 1.79 Catalyst-dependent chemoselective oxidation of β-ionone **137**

the reaction. Smaller dibenzyl diselenide **1o** is sterically less hindered than diselenides with a second substituted carbon, considering the nucleophilic attack and, because it is electron-enriched compared to diaryl diselenides, the seleninic acid **122o** derivative presents a good nucleophilicity (Scheme 1.80B).

In their studies on the properties of new glycerol-based solvents in oxidation reactions, García and Arends have used the epoxidation of cyclooctene with H_2O_2 50% (4 equiv.) catalyzed by bis[3,5-bis(trifluoromethyl)phenyl] diselenide **1s** (1 mol%) [118]. A detailed analysis of solvent polarity parameters indicated a quantitative relationship between solvent polarity and the rate of the epoxidation reaction. The authors observed that solvents with high hydrogen bond (HB) donor ability, and low HB acceptors are the best ones for this reaction. Several of the glycerol-based solvents have met these requirements, while 2,2,2-trifluoroethanol was the best among all the tested ones.

1.5.1.5 Synthesis of 1,2-Diols from Alkenes

Santi and coworkers described for the first time the synthesis of 1,2-diols **138** from olefins **54** catalyzed by $(PhSe)_2$ **1a** using H_2O_2 as the oxidant in a 3:1 mixture of H_2O and MeCN as the solvent [46]. The authors used eight different cyclic and acyclic olefins **54** and performed the reaction using 20 mol% or a stoichiometric amount of **1a**. In some cases, the reaction time was drastically reduced using a stoichiometric amount of **1a**, however the high diastereoselectivity observed in the reaction was not a function of the amount of the catalyst. Yields from 25 to 100% were obtained after 24–300 h of reaction at room temperature (Scheme 1.81).

As it can be seen in Scheme 1.81, *trans*-4-octene afforded exclusively *anti*-octane-4,5-diol **138c**, while 1-methyl-1-cyclohexene and 1-phenyl-1-cyclohexene afforded preferentially the *syn*-1,2-diols in good selectivity. Based on these results and ^{77}Se-NMR studies, the authors have proposed a mechanism to explain the *syn*/*anti* selectivity in the reaction (Scheme 1.82). The reaction passes by the formation of epoxide **A**, formed by the oxygen-transfer of benzeneperseleninic acid **122a** to alkene **54**. Following, intermediate **A** could rapidly react with water by a S_N2 reaction to form *anti*-**138** (path B) or through a carbocation intermediate **B**

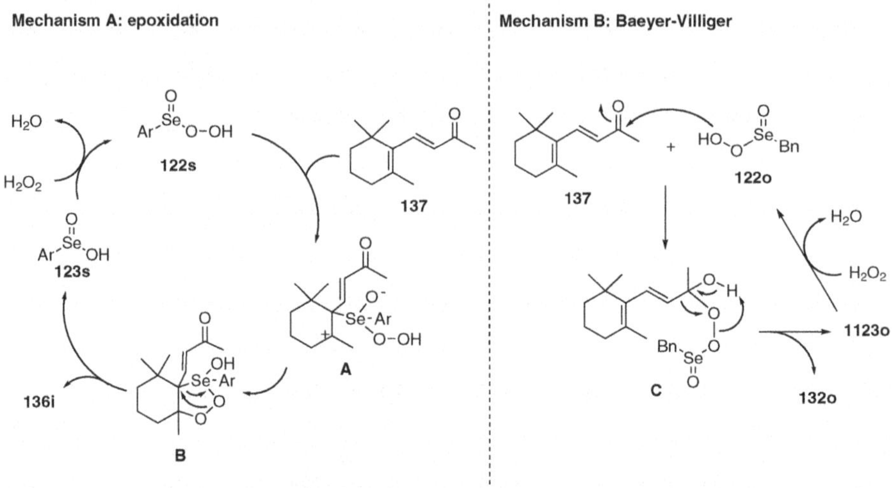

Scheme 1.80 Mechanism of chemoselective oxidation of β-ionone **137**

(path A), according the electronic and steric properties of the alkene. Because 1-methyl-cyclohexene and 1-phenyl-cyclohexene generate stable carbocations, they react preferable by path A, affording the *syn*-diols **138d** and **138e** as the major products. In the case of (+)-*p*-menth-1-ene, however, the contradictory preference for *anti*-**138a** is a consequence of the anchimeric assistance of the other substituents in the cyclohexyl ring.

Santi and coworkers developed the synthesis of *anti*-1,2-diols **138** and β-methoxyalcohols **139** using L-selenocystine **1u** as a pre-catalyst in a GPx-inspired mechanism [45]. The authors have realized that in the absence of a cofactor (GSH), amino acid **1u** could be successively oxidized to the respective perseleninic acid **122u**, which is responsible to transfer an oxygen to the double bond of alkene **54** as showed for **122a** in Scheme 1.82. In fact, (+)-*p*-menth-1-ene was selectively converted to *anti*-**138a** after 168 h in 80% yield and without racemization in the presence of 2 mol% of **1u** and 4 equiv. of H_2O_2 30% in water. The optimized conditions were extended to several other alkenes and a total of seven alkenes were efficiently converted to the respective diols in 45–80% yields and up 87% *ee* (Scheme 1.83). The authors used also chiral diselenides **1v** and **1x** to investigate the stereoselectivity of the reaction to prepare 1,2-diols **138d–g**. Larger amounts of catalysts **1v** and **1x** (10 mol%) and of H_2O_2 (40 equiv.) were necessary to afford similar selectivities. The reaction rates, however, were drastically reduced using diaryl diselenides instead L-selenocystine **1u**. The water/catalyst **1u** system was reused up to five times with satisfactory yields, just by adding more alkene and 1 equiv. of H_2O_2 after each reaction.

Using methanol instead water as a co-solvent [45], the hydroxymehylation reaction occurs, affording the respective β-methoxyalcohols **139** in good to excellent yields using 1 mol% of L-selenocystine **1u** as the catalyst and 4 equiv. of H_2O_2 as the

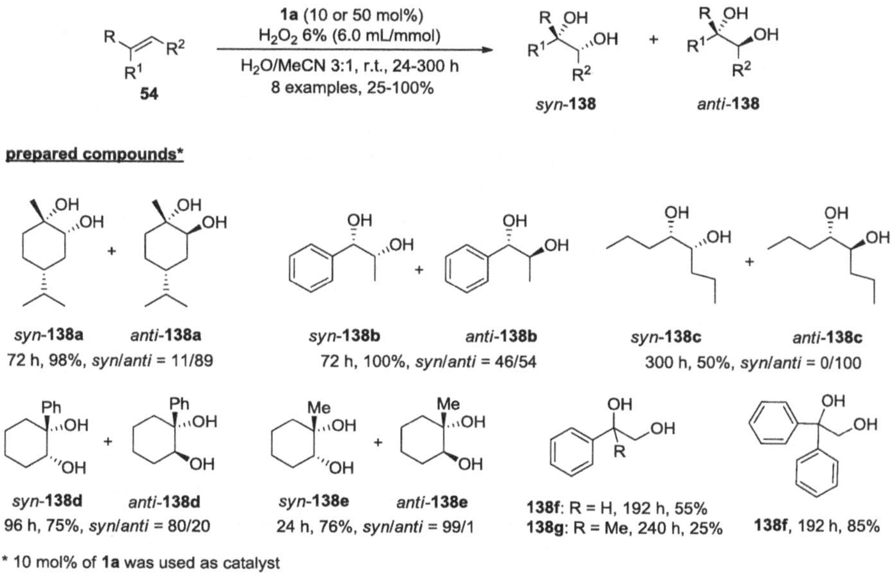

Scheme 1.81 Se-catalyzed synthesis of diols

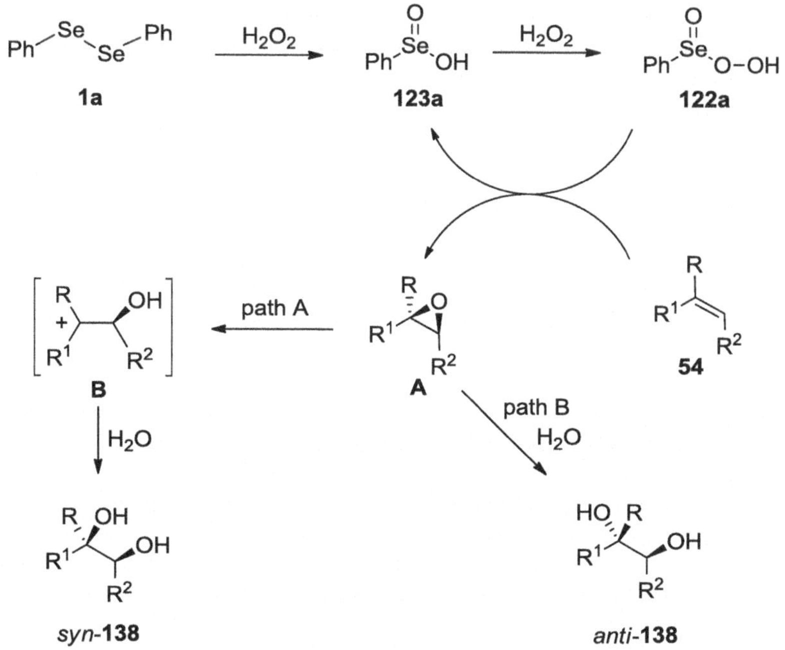

Scheme 1.82 Mechanism of Se-catalyzed synthesis of diols

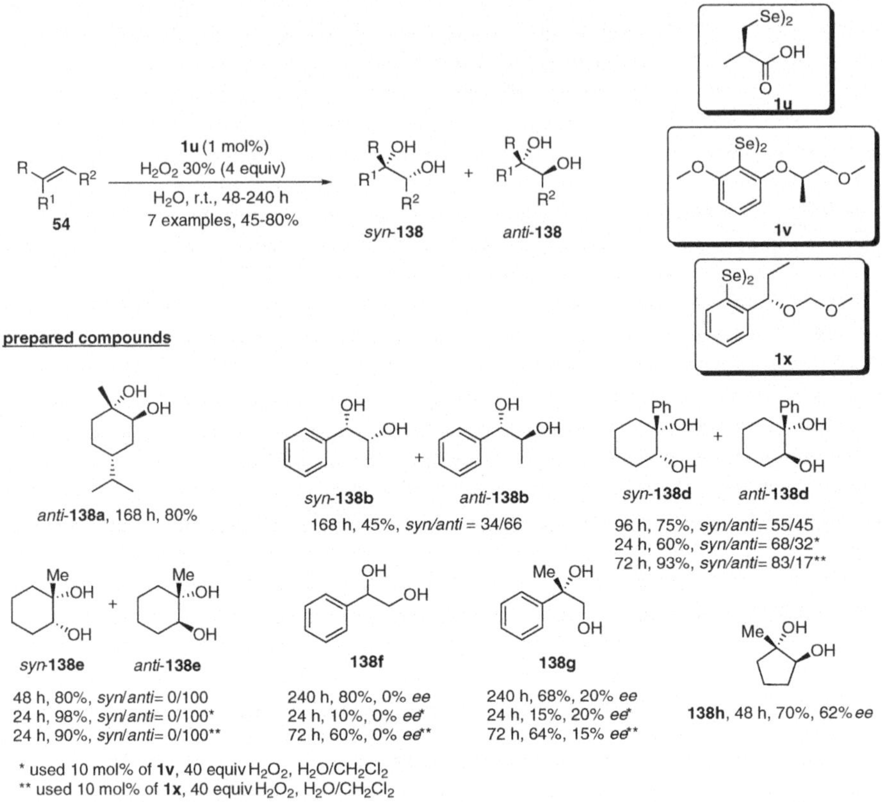

Scheme 1.83 Stereoselective Se-catalyzed synthesis of diols

oxidant. Because of a competitive noncatalyzed reaction, the hydroxymethylation reaction presented a considerably reduced stereoselective compared to the dihydroxylation one, despite the reactivity using methanol was increased (Scheme 1.84).

1.5.1.6 Oxocyclization of Functionalized Alkenes

If the nucleophile is appended to the alkene, as in the case of alkenoic acids **34** and alcohols **88**, products of oxocyclization are obtained, like the very useful hydroxy-functionalized γ-butyrolactones **140**, tetrahydrofurans **141**, and tetrahydropyran **142**. This strategy was firstly demonstrated by Santi and coworkers [119], who used $(PhSe)_2$ **1a** (5 mol%) as the pre-catalyst in the presence of H_2O_2 30 wt% (4 equiv.) to promote a one-pot epoxidation/ring-opening reaction of α,β- and γ,δ-unsaturated acids and alcohols, leading to the *O*-heterocycles **140–142** in 60–90% yields after 8 h at room temperature (Scheme 1.85). When the chiral pre-catalyst dimenthyl diselenide **1y** was used, (*E*)-hexenoic acid was converted to the corresponding

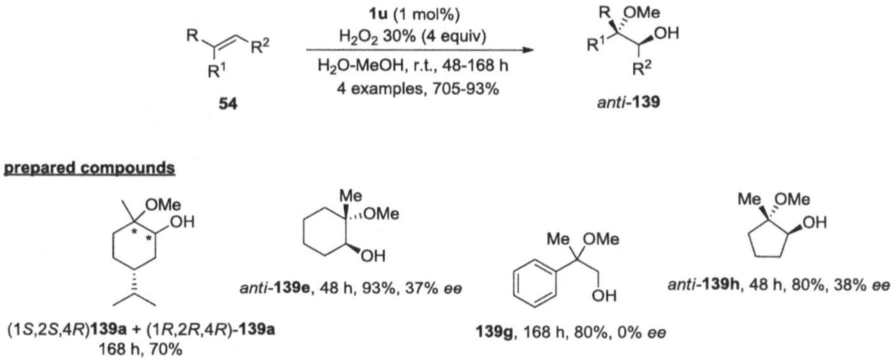

Scheme 1.84 Oxidative functionalization of olefins

γ-butyrolactone **140a** in 95% yield and 28% *ee* under the optimized conditions (8 h at room temperature). The reaction mechanism is similar to that of Scheme 1.82, except that the nucleophilic species is appended to the alkene. The first step is the oxidation of diselenide **1a** or **1y** by H_2O_2 to form seleninic acid **123** and then perseleninic acid **122**, which then transfers an oxygen to the double bond to form an epoxide intermediate. Once formed, the epoxide undergoes an intramolecular nucleophilic attack from the oxygen of the acid or the alcohol, giving the respective heterocycles **140–142**.

Inspired by the excellent results obtained in the Se-catalyzed oxocyclization of alkenoic acids and alcohols, Santi and Gioiello designed a flow set-up that was successfully used to prepare hydroxy-functionalized lactones from alkenoic acids by a continuous-flow synthesis [120]. The authors used benzeneseleninic acid **123a** (10 mol%) as the pre-catalyst, which was oxidized in situ to benzeneperseleninic acid **122a** by H_2O_2. In a typical procedure, a solution of the alkenoic acid in AcOEt and a solution of the pre-catalyst **123a** (10–50 mol%), HCO_2H (10 mol%) and H_2O_2 30 wt% (5 equiv.) in H_2O/acetone were injected in the loops L, mixed in a T-piece and passed through the reactor R at room temperature for 50 min (Scheme 1.86). The authors have accessed 16 different lactones in 67–92% yields, including 5-hydroxy-4-decanolide **140i** (obtained in 82% yield), the sex pheromone of the parasitic wasp genus *Nanosia*. By using an integrated flow system with four reactors operating in parallel, 22 mmol (74% yield) of the natural-occurring 3-hydroxy-γ-butyrolactone **140g** was continuously prepared in high purity and with a productivity corresponding to 6 g/day.

The flow set-up used in this green protocol was equipped with loop injection systems, two HPLC pumps, a reactor coil (10 ml), a back pressure regulator (BPR, 100 psi) and a fraction collector FC (Fig. 1.6). A continuous liquid-liquid membrane-based separator was used to separate the aqueous phase from the organic solution. The recovering of the catalyst and unreacted compounds was successfully performed after the reaction, by washing the Amberlyst/silica column used to trap the Se-catalyst and starting materials with ethanolic NH_4OH (5% v/v).

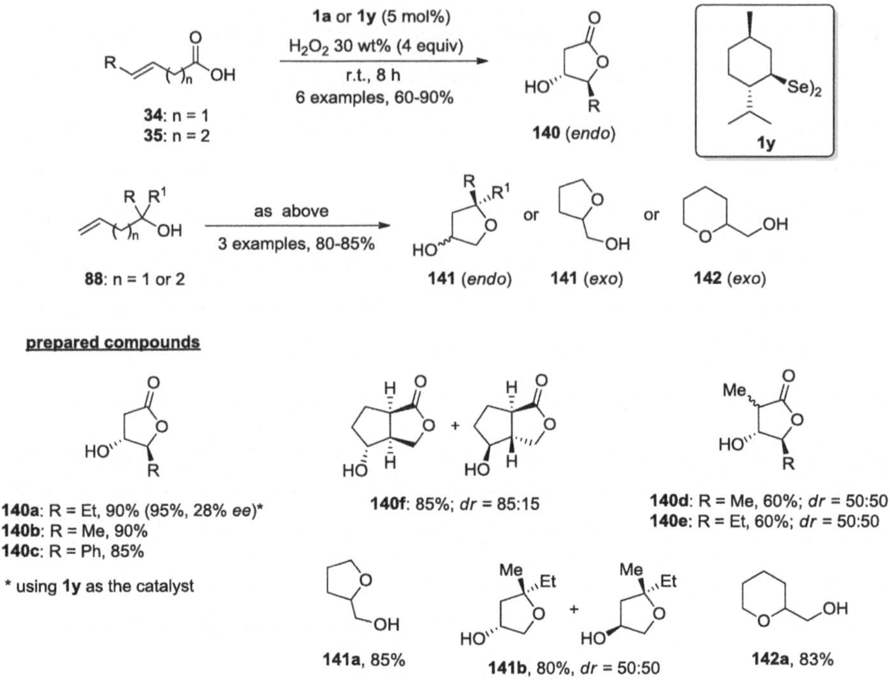

Scheme 1.85 Oxocyclization of functionalized alkenes

1.5.2 Hydrogen Peroxide in Halogenation Reactions

As mentioned before, chemists have been working to design synthetic catalysts with similar activity than natural ones, making possible to use H_2O_2 in a similar way to nature does. Concerning organoselenides, it has been shown through this chapter that the catalytic activity can be modulated by changing the structure of the compound, e.g., changing electronic and steric effects according the substituents in the selenium atom. Another approach used to improve the catalytic activity of chalcogen-based catalysts is through statistical means, by the synthesis of functionalized dendrimers [121–123]. The catalytic activity in dendrimers is a function both of the number of chalcogen atoms in the molecule and the dendrimer effect, in which the catalytic activity of individual groups increases or decreases with the number of generated dendrimers [124]. Detty and coworkers have designed and synthesized dendrimeric polyphenylselenides **146** which were used as catalysts in the oxidation of bromide with H_2O_2 (Fig. 1.7) [121, 122].

In its seminal work, Detty used the bromination of cyclohexene **54a** as a model to verify the catalytic activity of dendrimers **146a–d**, with 1, 3, 6, and 9 phenylselanyl groups respectively, and compared their activity with that of selenide **147** [121]. Because the low solubility of **146a–d** in water, the reactions were conducted in a biphasic system of CH_2Cl_2 and phosphate buffer (pH 6). The reaction was a good

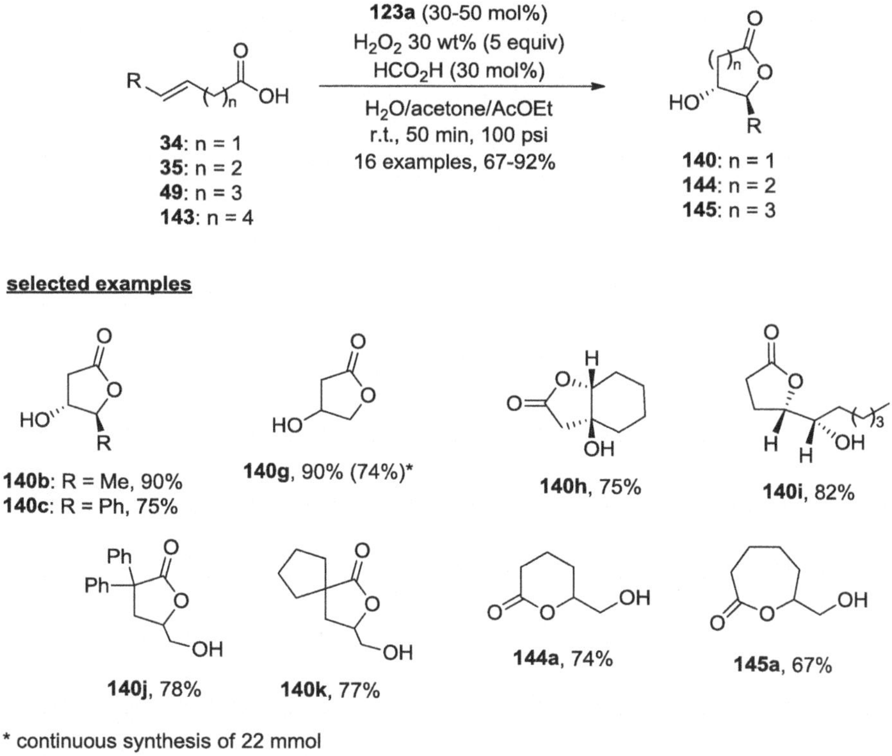

Scheme 1.86 Synthesis of lactones through flow system

indicator of the formation of electrophilic bromine species (NaOBr, Br_2, or Br_3^-), once cyclohexene was rapidly converted to *trans*-1,2-dibromocyclohexane **148** and *trans*-2-bromocyclohexanol **149** as the only products (Scheme 1.87).

The authors have prepared dendrimers containing sulfur, selenium and tellurium and compared their catalytic activity. The sulfur-derivatives were not able to activate H_2O_2 in the oxidation of bromide, while the tellurides presented the expected statistical effects in accelerating the reaction, with the relative initial rate [K_{rel}/PhTe group] increasing by a constant factor of 1, 3, 6, and 12. With the selenium analogues **146a–d**, however, an unexpected strong "dendrimer effect" or cooperativity was observed, and K_{rel}/PhSe group increased by a factor of 80 from monoselenylated **146a** to dendrimer **146b** (12 PhSe groups). On the basis of consumed H_2O_2, the turnover number for **146b** was >60 kmol of H_2O_2/mol of catalyst.

The large dendrimer effect observed for **146b–c** was not present when the selenoxide analogue was used as the catalyst. The authors performed a detailed study to explain the origin of the dendrimer effect in **146b–c** and they discovered that the primary oxidant for converting selenide to selenoxide **A** in dendrimers is "Br^+," initially produced by the uncatalyzed background reaction of H_2O_2 with NaBr. The large dendrimer effect observed with increasing number of

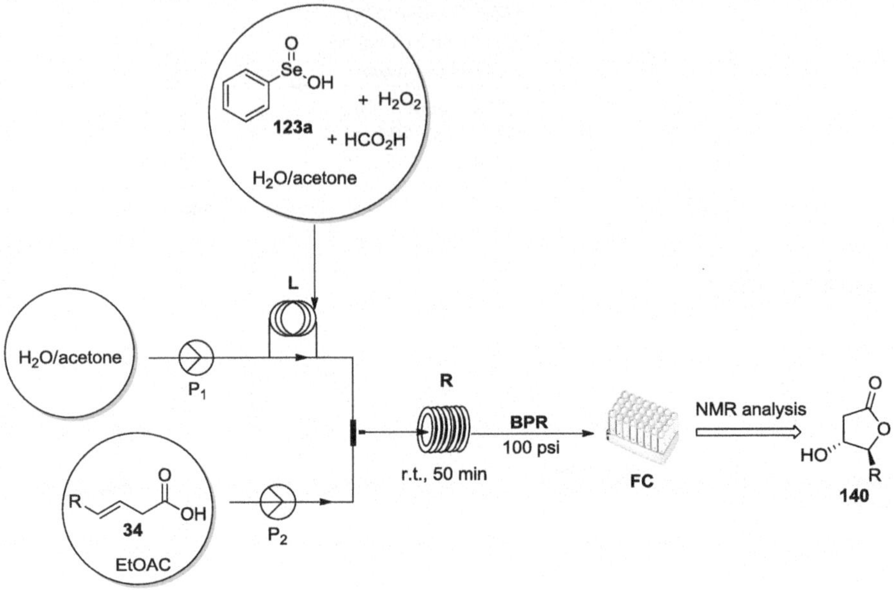

Fig. 1.6 Flow set-up used to prepare hydroxy-lactones

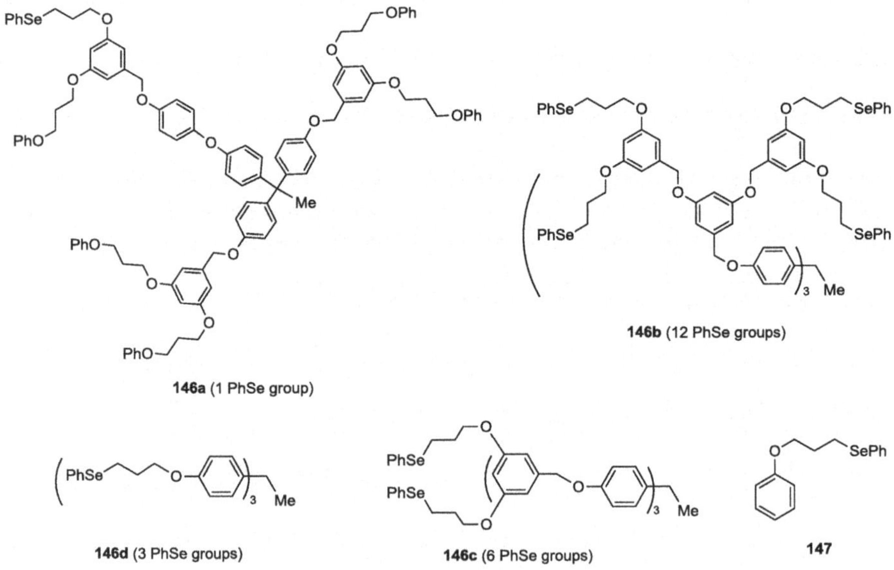

Fig. 1.7 Dendrimeric polyphenylselenides **146** and selenide **147**

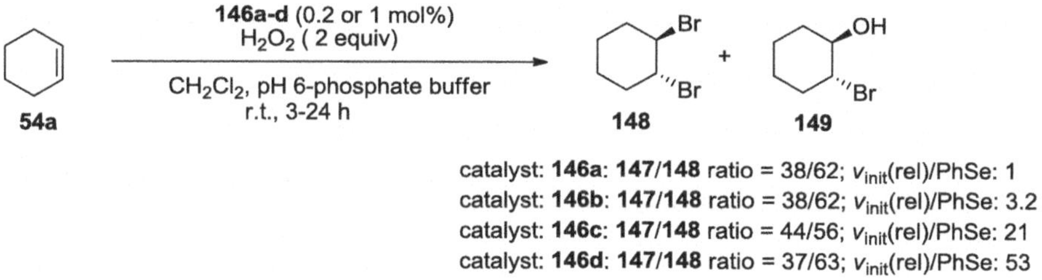

Scheme 1.87 Bromination of cyclohexene

PhSe groups is caused by the autocatalysis due to a growth in the rate of oxidation caused by the development of new catalytic groups (selenoxide ones) (Scheme 1.88).

In their studies on the substituent effects in the arylseleninic acid-catalyzed bromination of alkenyl acids, alkenyl alcohols, and electron-rich arenes using NaBr and H_2O_2, Detty and coworkers evaluated six differently functionalized seleninic acids [125]. By this procedure, unsaturated acids **34** and **45** afforded the respective bromolactones **69** in 55–84% yields after 20–24 h at room temperature. When 4-pentenoic acid was used, besides lactone **69h**, 1,2-dibrominated acid **150** was isolated in 16% yield, while β-cyclohexenyl acetic acid produced a mixture of γ- and β-lactones, with predominance of the second (64% yield) after 24 h of reaction. Alkene 4-penten-1-ol **88i** delivered 2-bromomethyltetrahydrofuran **151i** in 17% yield together with 4,5-dibromopentan-1-ol **152i** (48%) after 24 h of reaction (Scheme 1.89). This biphasic system was successfully used in the bromination of electron-rich arenes **153**, giving 82–86% yields of the mono-brominated derivatives **154** after 19–24 h of reaction.

The authors discovered that unsubstituted benzeneseleninic acid **123a** and 4-methoxyphenylseleninic acid **123z**, were the most active catalysts. This observation is consistent with a mechanism that involves a direct attack of a bromide anion at the Se-O bond of perseleninic acid **129** (Scheme 1.90). The first step in the reaction is the oxidation of **123** to **122** by H_2O_2, which undergoes a nucleophilic attack by the Br^- to form intermediate **B** and OH^-. Then, a second bromide anion reacts with **B** to form Br_2 and regenerates benzeneseleninic acid **123** for a new cycle. Alkenes or electron-rich arenes reacts with the bromine species generated in situ from the reaction of **B** with Br^-.

The same bromination reactions described in Scheme 1.91 were performed using selenoxides **155** as catalysts instead arylseleninic acids **122** [125]. After a screening through different arylbenzylselenoxides, the authors have found that the presence of electron-releasing groups at the aromatic ring directly bonded to selenium accelerate the bromination reaction of 4-pentenoic acid **45h**. When 2.5 mol% of 2-[(dimethylamino)methyl]phenyl selenoxide **155a** was used as a catalyst in the presence of NaBr and H_2O_2 in the presence of CH_2Cl_2 and phosphate buffer (pH 6), **45h** was converted to the respective bromolactone **69h** in 93% yield after 8 h at room temperature. The reaction was faster and the yield higher than that using arylseleninic

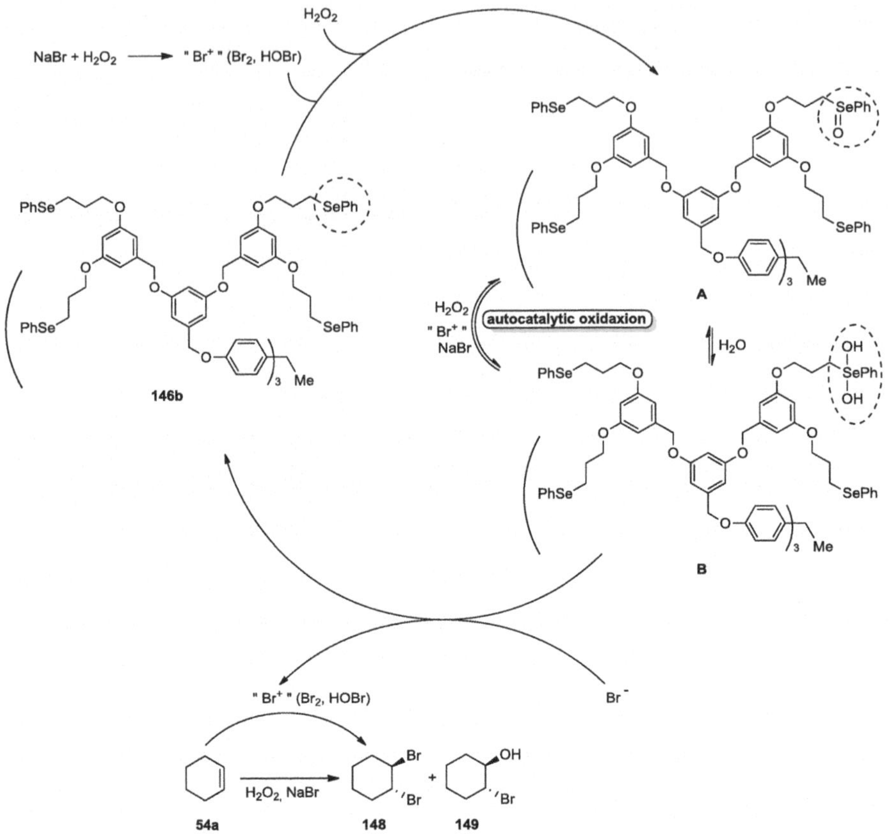

Scheme 1.88 Bromination of olefins catalyzed by Se-containing dendrimers

acids **122** as the catalyst (compare with Scheme 1.84). In the bromination of electron-rich arenes **159**, however, the catalytic activity of **155a** was slightly inferior to that of **122a**, with an increased preference for the 4-bromo-substituted anilines **154b** and **154e** (Scheme 1.91).

The structure–catalytic activity studies with different selenoxides **155** showed that the presence of substituents in the aromatic ring affect the basicity of the selenium atom and the stability of its conjugated acid **A**. Arylbenzylselenoxide **155a**, with a dimethylamino substituent, is more effective in stabilizing the conjugated acid **A**, through the formation of chelate **D**, providing additional stability to the hydroxyselenonium intermediate **A**. This observation is consistent with the involvement of the hydroxy perhydroxy selenane **156**, the effective oxidant agent, which is formed by the reaction of intermediate **A** with H_2O_2. According to the authors, after formation of **156**, the oxidation of bromide to electrophilic bromine could follow two mechanism steps: (a) the direct attack of Br^- at the -OH to give HOBr, regenerating selenoxide **155** by releasing hydroxide, or (b) a nucleophilic attack of Br^- at

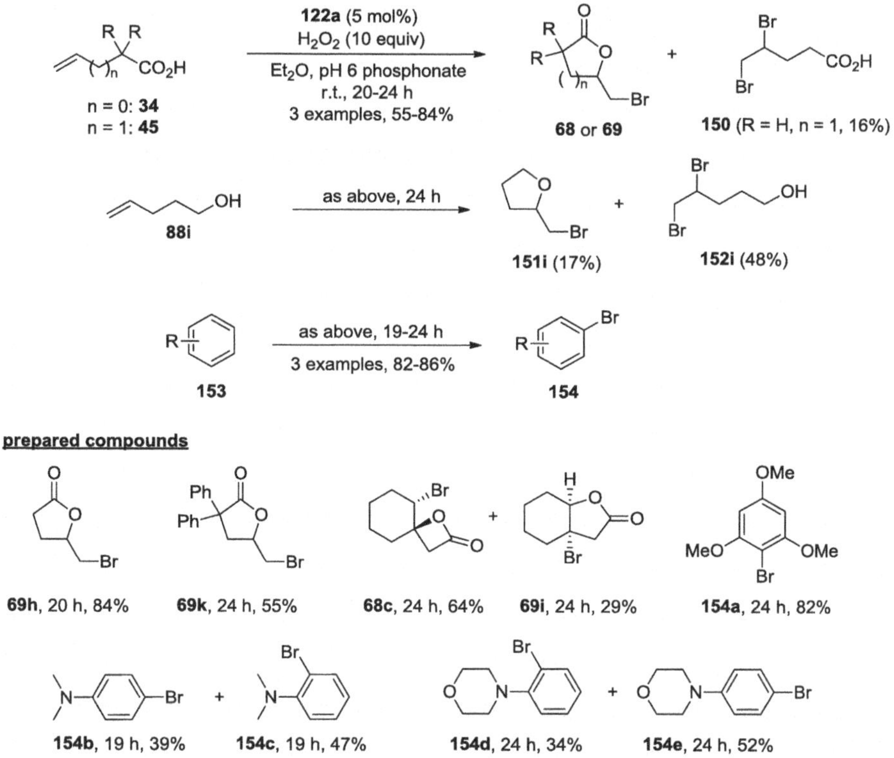

Scheme 1.89 Bromination of alkenyl acids, alkenyl alcohols and electron-rich arenes

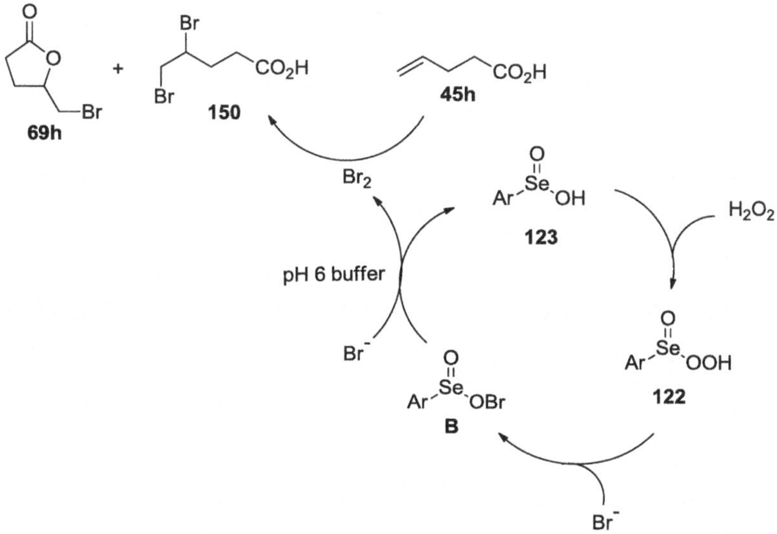

Scheme 1.90 Mechanism of oxidative bromination reaction catalyzed by **123**

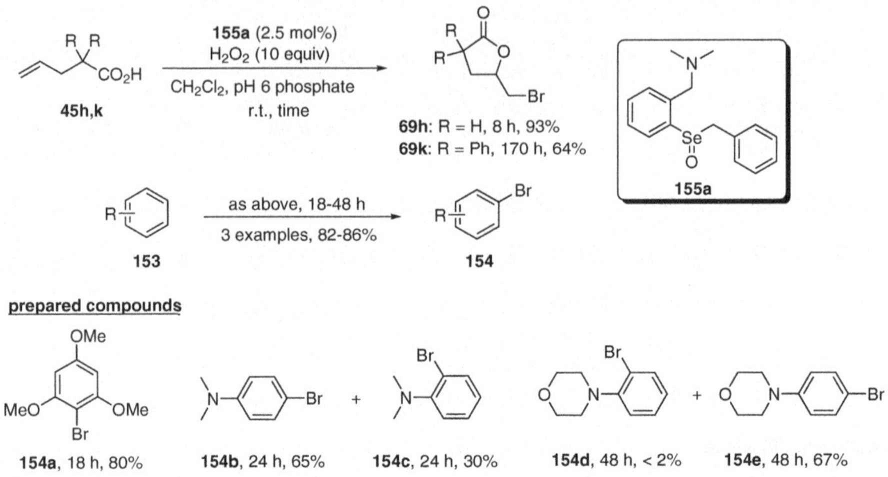

Scheme 1.91 Bromination of alkenoic acid and aromatic compounds using selenoxides **155** as catalysts

the Se-O oxygen atom of **156** to form compound **B**, that could be itself the brominating agent or could react with Br^- at the O-Br bromine atom to produce molecular bromine (Scheme 1.92).

Alternatively, the selenoxide catalyst was sequestered in a halide-permeable xerogel (a 10/90 mixture of 3-aminopropyltriethoxysilane/tetraethoxysilane—APTES/TEOS), affording a 23-fold more active catalyst compared to the xerogel-free one [126]. To make possible the covalent attachment of the selenoxide within the xerogel, selenoxide **155b**, containing the benzyl alcohol group, was previously prepared and allowed to react with the xerogel under sonication for 1 h. The amount of selenoxide recovered was inferior do 5%, indicating that >95% of **155b** was sequestered in the xerogel. This new Se-catalyst was successfully used in the bromination of alkenoic acid and aromatic compounds using $NaBr/H_2O_2$ in a 1:1 mixture of dioxane and pH 6 phosphate buffer. The main advantage in using the sequestered catalyst was the easy separation after reaction completion by simple filtration and washing with water and ether. Despite the selenium amount of APTES/TEOS-**155b** has remained essentially the same after up five reuses, the catalytic activity gradually diminished and only 65% yield of bromolactone **69h** was obtained in the fifth reaction from pentenoic acid **45h** (Scheme 1.93). The authors have proposed for the catalytic activity of APTES/TEOS-**155b** the same mechanism described in Scheme 1.88, for the xerogel-free selenoxide **155**.

More recently, Detty and coworkers prepared eight TEOS xerogel-sequestered organochalcogen compounds, including diorganyl tellurides and selenides, and diorganyl diselenide bearing triethoxysilane functionality [127]. The reaction of pentenoic acid **45h** to produce bromolactone **69h** was used as a model to verify the catalytic activity of the TEOS-catalysts and all the sequestered chalcogen showed activity, accelerating the reaction of $NaBr/H_2O_2$ to produce electrophilic bromine.

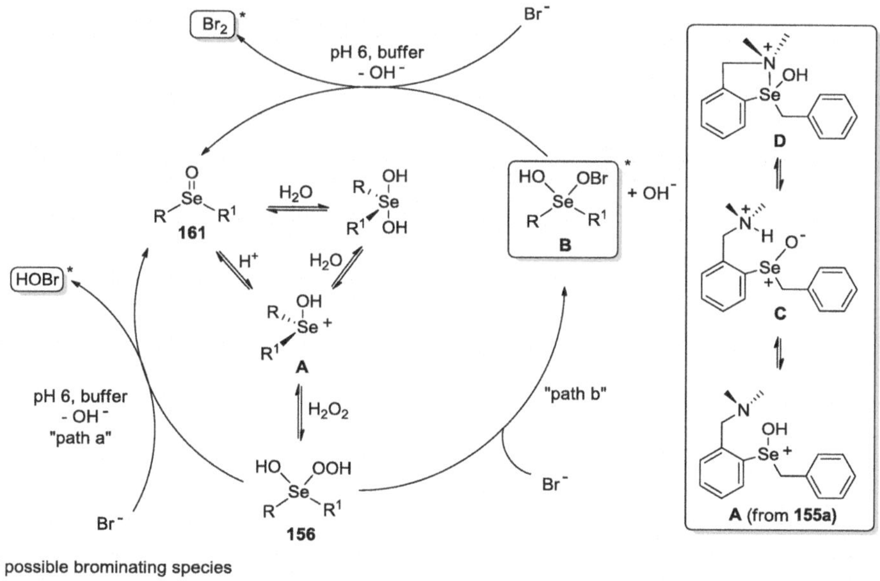

Scheme 1.92 Mechanism of oxidative bromination reaction

The best results were obtained using trimethoxy(4-(phenyltelluranyl)phenyl)silane and 2,2′-bis(*N*-(3-(triethoxysilyl)propyl)benzamide diselenide **1y** as pre-catalysts. In the recycling studies, diselenide **1y** presented the best results, and this performance was attributed to the oxidation of diselenide xerogel TEOS-**1y** by H_2O_2 would generate 2 equiv. of xerogel-sequestered seleninic acid TEOS-**123y**. Because seleninic acids do not undergo *syn*-elimination, it remains sequestered in the TEOS and consequently, the rates of bromination did not change from the initial reaction through the reuse (Scheme 1.94).

In a collaborative work, Braga and Detty designed three new water-soluble imidazolium-containing dibenzyl diselenides and tested them as catalysts in the bromination of alkenoic acids and electron-rich aromatic compounds using NaBr/ H_2O_2 system [128]. Among the new diselenides used, 3-methyl-imidazolium substituted **1z** was the more active, being more effective than diphenyl diselenide **1a** and ionic liquid 1-benzyl-3-methylimidazolium bromide **157** in the activation of H_2O_2, indicating that a synergism could occur when these two groups are connected in the same molecule. When comparing the reaction times of the bromination catalyzed by dendrimeric, xerogel-sequestered or isolated selenoxides, described above, the imidazolium diselenide **1z** was more active, affording the expected products in only 2.5–7 h of reaction at room temperature. Despite the reaction works well using only an aqueous phosphate buffer (pH 4.2) as the reaction media, the addition of dioxane (3:1 mixture) has a beneficial effect, reducing the reaction time and increasing the product yields. The higher activity of **1z** compared to the other imidazolium-containing diselenides was explained, at least in part, by the electrostatically stabilization promoted by the *ortho* positioned

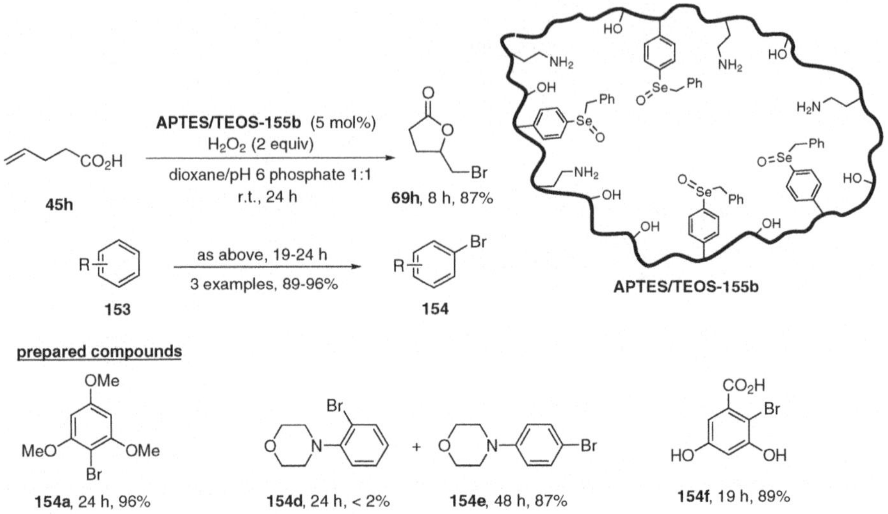

Scheme 1.93 Bromination of alkenoic acid and aromatic compounds catalyzed by Se-containing halide-permeable xerogel

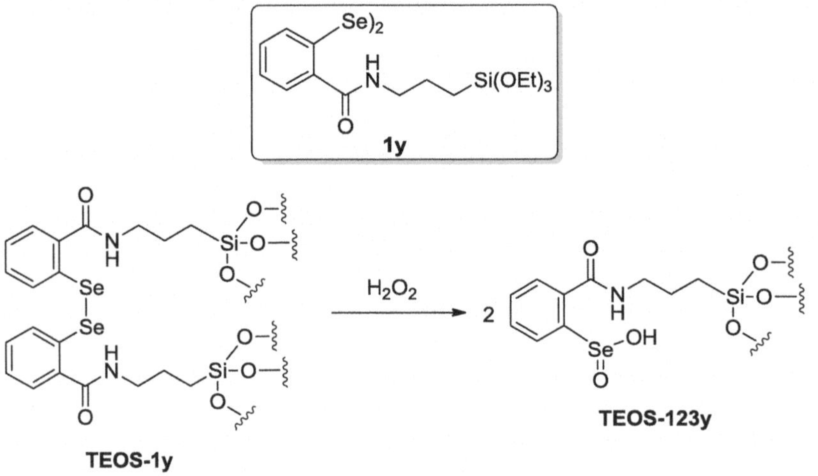

Scheme 1.94 Synthesis of xerogel-sequestered seleninic acid

imidazolium ring, which increases the stability of the negative charge developed in the perseleninic acid **122z** (Scheme 1.95).

New discoveries on the mechanism and intermediates involved in the oxidation of phenylselenides by H_2O_2 were recently published by Bellanda, Orian, and coworkers [129]. The authors have combined ^{1}H and ^{77}Se-NMR analysis with computational calculations, to observe the intermediates and products of the reaction of four organoselenium compounds with H_2O_2. Phenyl butylselenide **71f**, bis(phenylselanyl)methane **158**, 1,2-bis(phenylselanyl)ethane **159** and diphenyl diselenide (**1a**) were subjected to the reaction with aqueous H_2O_2 in a NMR tube so the changes in chemical shifts in ^{1}H and ^{77}Se-NMR spectra were used to follow the reaction (Fig. 1.8).

The conclusions of this interesting study are outlined in Scheme 1.96. PhSeBu **71f** is directly oxidized by H_2O_2 to selenoxide **160**, with no intermediate species being observed in the NMR spectra. Once formed, selenoxide **160** can be oxidized by additional H_2O_2 to the hydroxy perhydroxy selenane **161** (Scheme 1.96a). Bis(phenylselanyl)methane **158** undergoes a stepwise double oxidation: reaction with 1 equiv. of H_2O_2 gives the mono oxidized species **162** that, by reaction with excess H_2O_2 (3 equiv.), is converted to a diastereoisomeric mixture of the di-selenoxides (*R*,*R*)-**163** and (*R*,*S*)-**163** (Scheme 1.96b). The oxidation of 1,2-bis(phenylselanyl)ethane **159** afforded a mixture of benzeneseleninic acid **123a** and phenylselenoethene **166**, which is derived from a *syn*-elimination reaction due to the presence of β-protons in **164**. Analysis of the kinetic profile of the reaction between **159** and H_2O_2 indicates the fast formation of the mono-oxidized species **164** and a slow formation of di-selenoxide **165** and benzeneseleninic acid **123a**. Following, **164** starts to disappear and the disappearance accelerates, giving rise to more **165**. However, a fast β-elimination in **165** results in a rapid formation

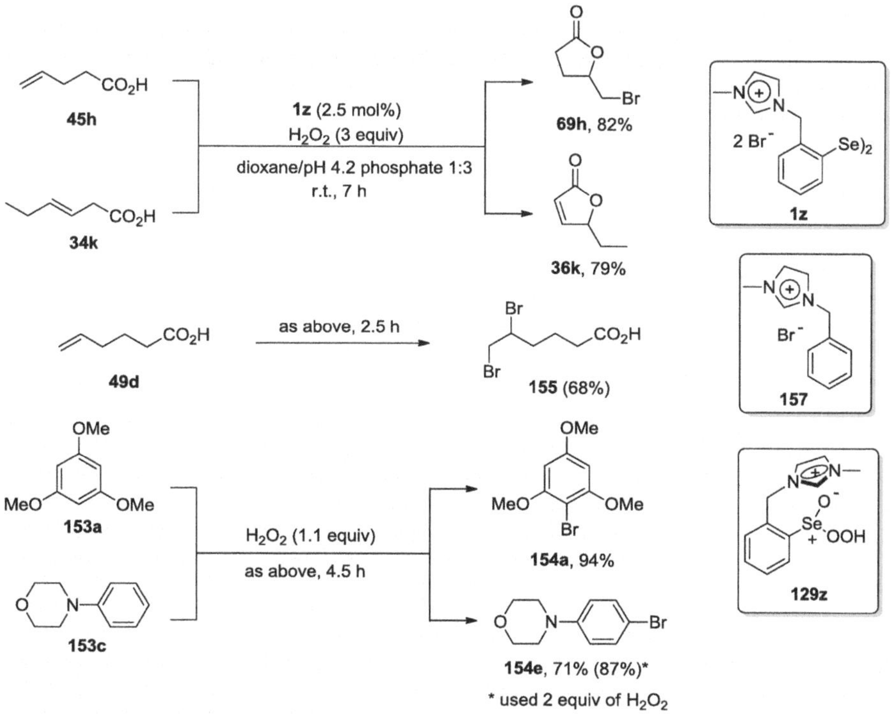

Scheme 1.95 Bromination of alkenoic acids catalyzed by imidazolium-containing dibenzyl diselenides

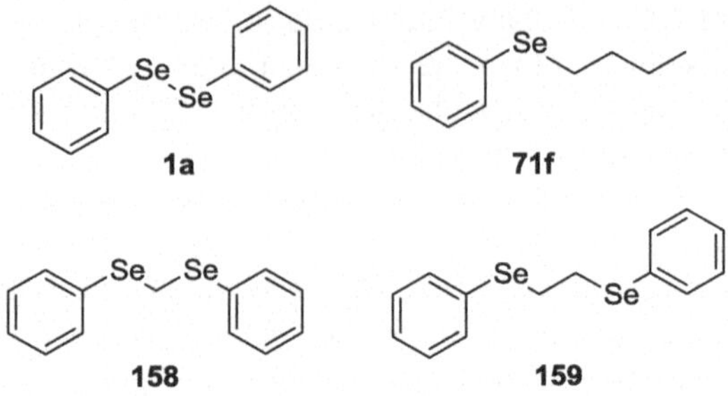

Fig. 1.8 Phenyl selenides studied by Bellanda et al. [129]

of vinyl selenoxide **167** and benzeneseleninic acid **123a**, that can further react to form perbenzeneseleninic acid **122a**. Once formed, **122a** can itself oxidize **164** and **165**, starting an autocatalytic process (Scheme 1.96c). The direct oxidation of diphenyl diselenide **1a** to benzeneseleninic acid **123a** by the Se-Se bond cleavage is a relatively slow process. However, as soon as some **123a** is formed, an acceleration of the reaction is observed and it is due the autocatalytic feature of the second step of the reaction (Scheme 1.96d).

In the presence of H_2O_2, **123a** is oxidized to perbenzeneseleninic acid **122a** that, similarly to described above for the oxidation of **159**, can catalyze the first step of the reaction of **1a** (formation of mono- and di-selenoxides), thus accelerating the whole process via an autocatalysis (Scheme 1.97). An additional conclusion of these experiments is that the oxidation by **122a** is energetically favored over the oxidation by H_2O_2. This explains the acceleration of the reactions with substrates **164** and **1a** (by autocatalysis), after a slow initial step (oxidation by H_2O_2).

1.6 Se-Catalyzed Radical Reactions

Crich and Sannigrahi developed a Se-catalyzed radical arylation of benzene with *ortho*-functionalized iodobenzenes **168**, affording aryl cyclohexadienes **169** by a reductive C-C bond forming sequence [130]. The active catalyst in this reaction is benzeneselenol, generated in situ from $(PhSe)_2$ **1a** and tributyltin hydride. The role of benzeneselenol is trapping the cyclohexadienyl radical intermediate, enabling the isolation of the desired compounds **169**. The aryl cyclohexadienes were easily cyclized with PhSeBr, delivering tetrahydrodibenzofuranes **170**, tetrahydrodibenzopyranone **171** or tetrahydrocarbazole **172**, according the functionality present in the starting iodobenzene **168**. In a typical procedure, degassed benzene is dropwise added to a refluxing mixture of iodobenzene **168**, $(PhSe)_2$ **1a** (20 mol%), azobisisobutironitrile (AIBN, 0.1 equiv.), and Bu_3SnH (1.2 equiv.) in benzene, giving the products after one additional hour of reflux. Important, the addition of benzene is dropwise, during around 12 h. By this procedure, several *o*-substituted iodobenzene derivatives were successfully coupled with benzene to

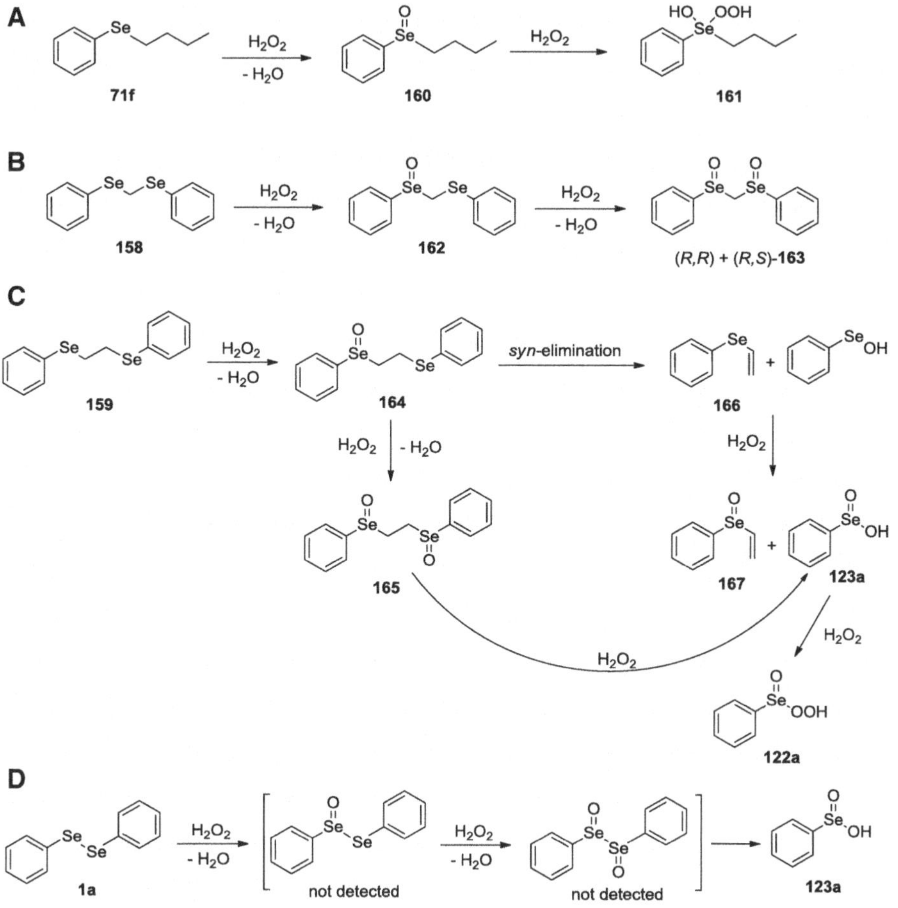

Scheme 1.96 (**A**) Sequential oxidation of PhSeBu **71f**; (**B**) Sequential oxidation of PhSeBu **158** to disteromeric **163**; (**C**) Oxidation and *syn*-elimination of **159**; (**D**) Oxidation of **1a** into **123a**

afford the respective aryl cyclohexadienes **169** in 41–54% yields, including compounds derived from *o*-iodophenol, *o*-iodobenzoic acid, *o*-iodovanilin, and *o*-methoxycarbamoyliodobenzene. Interestingly, 1-bromo-4-iodobenzene afforded exclusively 4-bromophenyl cyclohexadiene **169e**, indicating a greater reactivity for the C_{sp2}-I bond cleavage over the C_{sp2}-Br one (Scheme 1.98).

The radical arylation of benzene catalyzed by PhSeH was explored by Crich and Rumthao in the total synthesis of carbazomycin B **173**, an inhibitor of 5-lipoxygenase and antifungal compound [131]. The properly substituted *o*-methoxycarbamoyliodobenzene **168f** was subjected to the radical arylation conditions to afford the aryl cyclohexadiene **169f** in 40% yield. Cyclization of **169f** with PhSeBr delivered the tetrahydrocarbazole **172f**, which undergoes a selenoxide elimination-aromatization reaction using TBHP to afford the acylated carbazomycin, that is deprotected in the presence of hot NaOH to deliver the desired

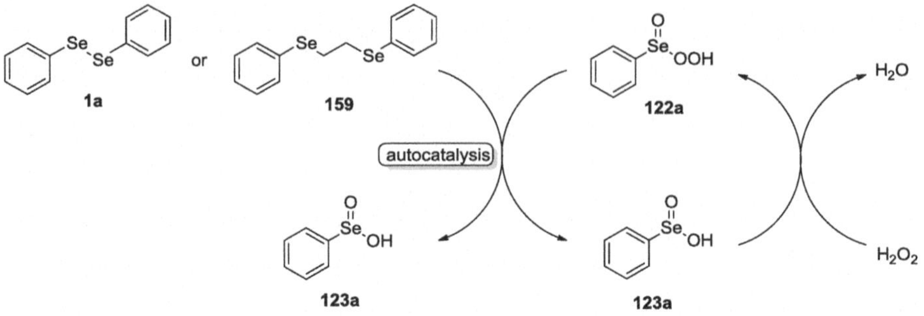

Scheme 1.97 Autocatalytic oxidation of **1a** or **159**

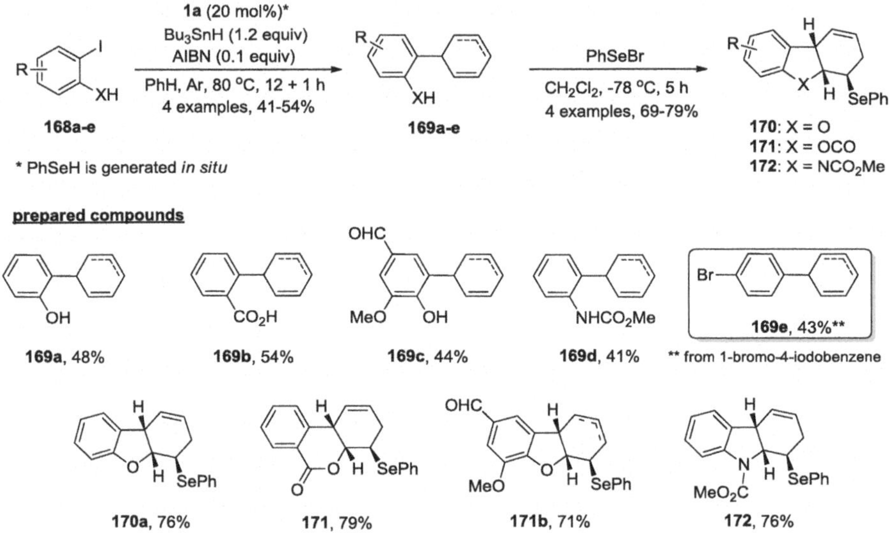

Scheme 1.98 Se-catalyzed radical arylation of iodobenzenes

carbazomycin B in an overall yield of 12% (four steps). This is a remarkable example of the versatility of organoselenium chemistry, both in catalysis and in stoichiometric reactions to prepare bioactive more complex molecules (Scheme 1.99a). The authors have also used this approach in the total synthesis of dibenzofuran peptide **175**, the Kelly's β-sheet initiator, starting from iodobenzene **168g** [132]. The key intermediate aryl cyclohexadiene **169g** was obtained in 61% yield and after eight steps it was converted to the target molecule **175** in an overall yield of 8.2% (Scheme 1.99b).

The Se-catalyzed radical arylation approach developed by Crich was successfully applied in the addition of iodobenzene to furan and thiophene [133]. After trapping with PhSeH, the adduct radicals afforded 2-aryl-2,3-dihydrofuranes **176** and thiophenes **177** and their 2,5-dihydro isomers **176′** and **177′**. When *o*-iodophenol **168a** and *o*-iodovanilin **168c** were used as starting aryl

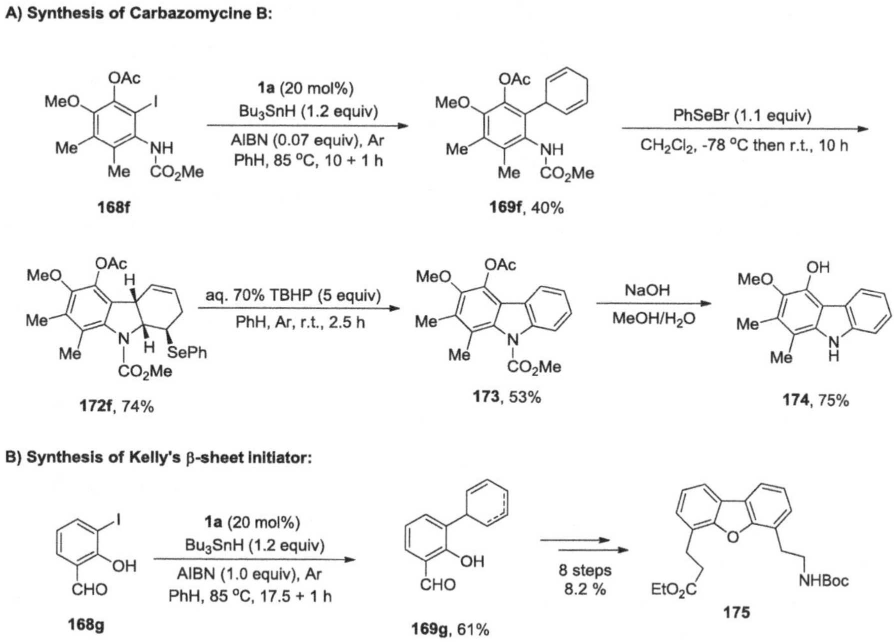

Scheme 1.99 (**A**) Total synthesis of carbazomycin B; (**B**) synthesis of the Kelly's β-sheet initiator

iodides, bridged bicyclic acetals **178** and **179** were obtained, through cyclization of the radical intermediates. Differently of the reactions using benzene, where AIBN was used as radical initiator, 2,2′-azobis(4-methoxy-2,4-dimethylvaleronitrile) (V-70; 0.2 equiv.) presented the best results in reactions with furan and thiophene. The reaction of *o*-iodophenol **168a** with pyrrole was also tested, however only 17% yield of the rearomatized 2-arylpyrrole **180** was obtained, together a complex mixture of co-products (Scheme 1.100). Despite a large excess of the heteroarene is used (around 30 ml/mmol iodobenzene), this is an interesting way to access the title compounds starting from simple reagents and in a selective way.

As showed in Scheme 1.100, reactions with thiophene gave preferentially 2-aryl-2,5-dihydro adducts, while furan derivatives gave preferentially the 2-aryl-2,3-dihydro ones. The difference in the regioselectivity of the hydrogen atom transfer from PhSeH to radical **B** in furan compared to thiophene one **B′** reflects the higher ability of alkylthio groups in localizing spin compared to the alkoxy ones. As a consequence of this spin localization and the reaction conditions, furan radical **B** has a preference for a distal quenching, forming 2,3-dihydro adduct **176** preferentially, while thiophene radical **B′** undergoes a proximal quench, favoring the 2,5-dihydro adduct **177′** (Scheme 1.101).

The ability of benzeneselenol in capturing radicals was explored by Clive and coworkers in the Se-catalyzed radical carbocyclization of *O*-trityloximes to prepare

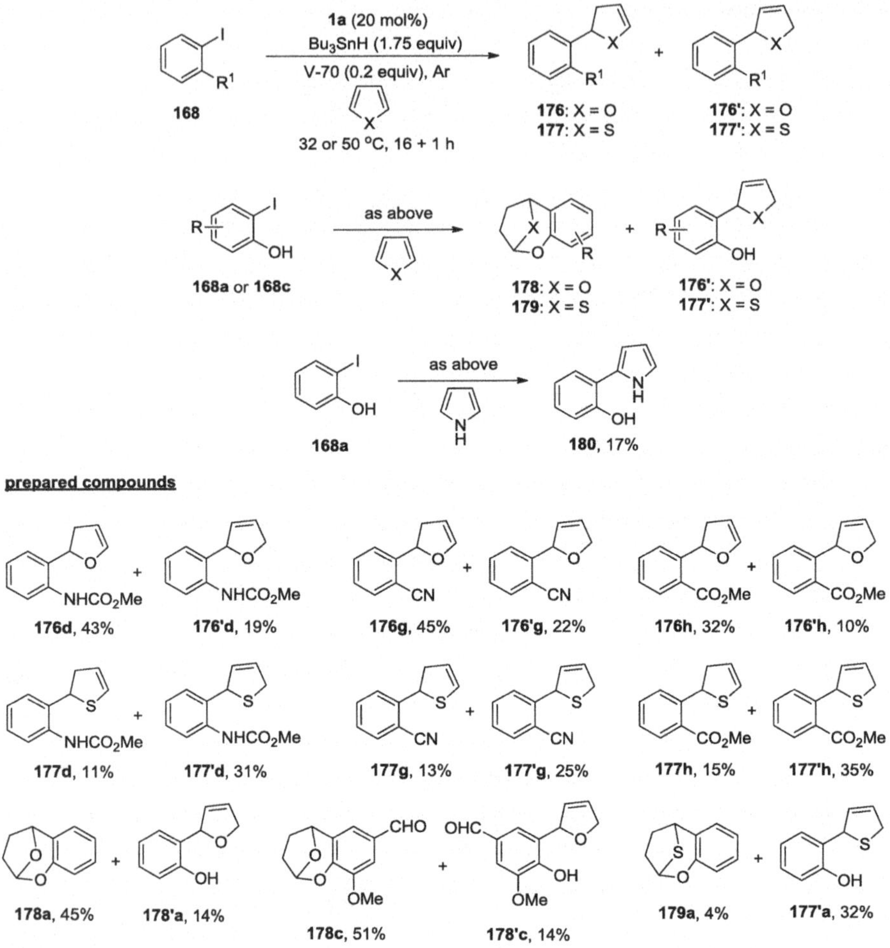

Scheme 1.100 Se-catalyzed radical arylation

five- and six-membered carbocyclic oximes [134]. According the authors, PhSeH rapidly capture the persistent triphenylmethyl radicals formed in the reaction, enabling the carbocyclization. In a typical reaction, a solution of Bu_3SnH (1.2 equiv.) and 1,1′-azobiz(cyclohexanecarbonitrile) (ABC; 1 equiv.) in THF was added dropwise (ca. 10 h) to a pre-heated (95 °C) mixture of iodo- or bromo-*O*-trityloxime **181**, $(PhSe)_2$ **1a** (20 mol%), and $^{i}Pr_2NEt$ (4 equiv.) in THF under argon. After stirring for additional 2 h, differently substituted cyclopentyloximes and cyclohexyloximes **182** and **183** were isolated in 56–94% yields, with worst yields being obtained using bromides instead iodides **181** (Scheme 1.102).

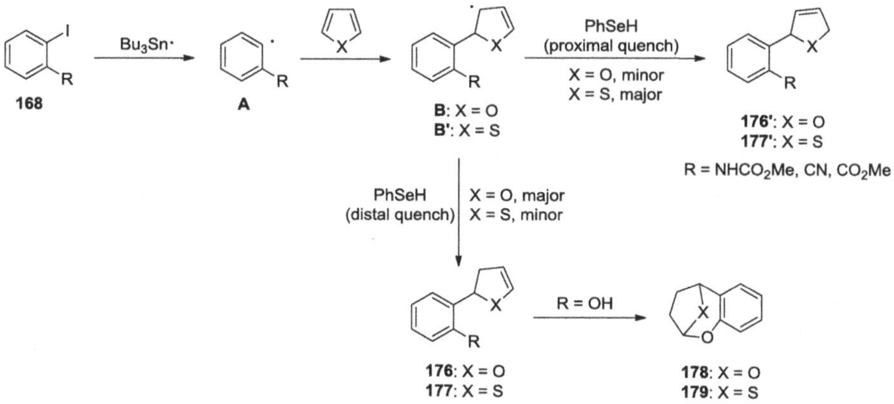

Scheme 1.101 Two different quenching in Se-catalyzed radical arylation

1.7 Miscellaneous Reactions

1.7.1 Se-Catalyzed Dehydration of Oximes

Yu and coworkers have prepared organonitriles **185** by a Se-catalyzed dehydration of aldoximes **184** by two different approaches, one using 3-fluorobenzeneselenenic acid **186** generated in situ [43] and other using previously prepared benzeneseleninic acid **123a** as the catalyst [135]. After a screening through 11 different diaryl and dialkyl diselenides **1**, the authors have found that bis(3-fluorophenyl) diselenide **1aa** was the best precatalyst for the reaction. In fact, the active catalyst was demonstrated to be the selenenic acid **186**, obtained by the reaction of **1aa** with equal amount of H_2O_2 [43]. The reaction was easily conducted simply by stirring a mixture of aldoxime **184**, diselenide **1aa** (4 mol%) and 30% w/w H_2O_2 (4 mol%) in acetonitrile at 65 °C for 24 h. By this strategy, alkyl, alkenyl, electron-rich, and electron-deficient aromatic aldoximes were efficiently converted to the respective nitriles **185** in 70–85% yields. However, the reaction failed with bulky mesityl aldoxime, which delivered mesityl aldehyde in 85% yield, and secondary and tertiary aliphatic aldoximes (^{i}Pr, cyclohexyl, ^{t}Bu), even using higher temperatures. The unreacted aldoximes were recovered (Scheme 1.103).

The recyclability of the organoselenium catalyst was showed in the dehydration of six different aldoximes, with good yields after up six reactions in 25 mmol scale, just by adding new portions of H_2O_2. A possible mechanism for the reaction involves the generation of unstable selenenic acid **186** by oxidation of diselenide **1aa** with H_2O_2, which self-condenses to selenenic anhydride **A**, with releasing of water. By reaction with the aldoxime **184**, intermediate **A** forms the mixed anhydride **B** that rearranges to its selenoxide **C**. After a selenoxide *syn*-elimination process, nitrile **185** is formed and selenenic acid **186** is regenerated for a new reaction (Scheme 1.104).

In spite the fact that hydrogen peroxide needs most of times an activation to be used as oxidant, it is a strong and unstable oxidant and can react with sensitive

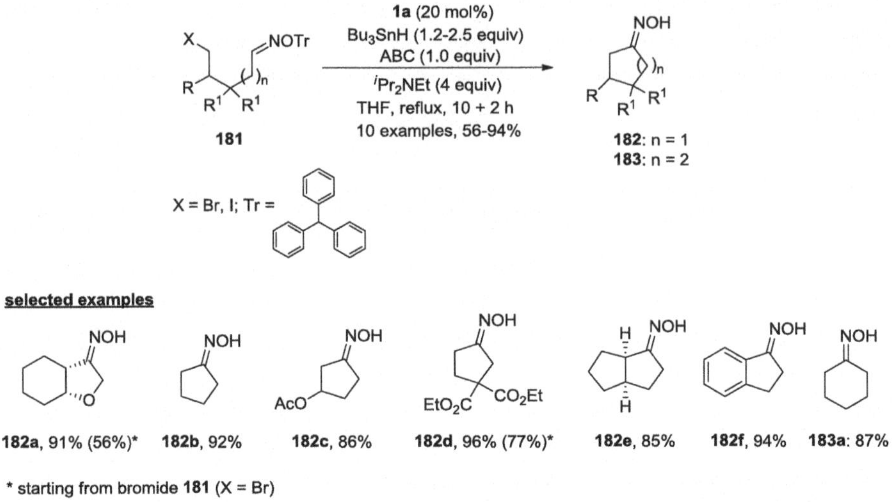

Scheme 1.102 Se-catalyzed radical carbocyclization of *O*-trityloximes

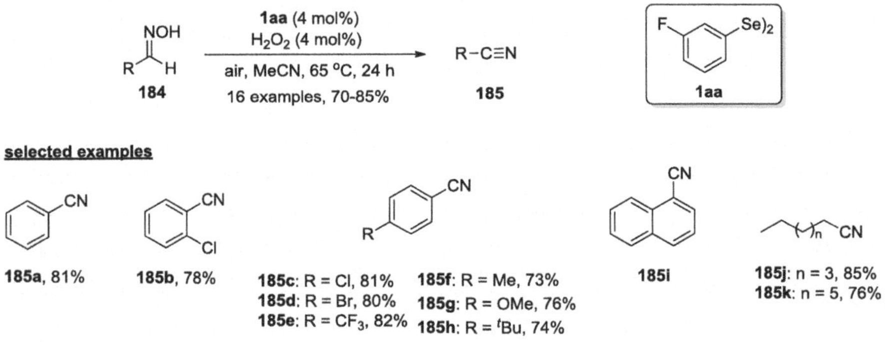

Scheme 1.103 3-Fluorobenzeneselenenic acid-catalyzed dehydration of aldoximes

groups in the substrate. In addition, it is difficult accurately weighing H_2O_2 used to prepare catalytic amounts of selenenic acid **186**, making the procedure described above not suitable for routine laboratory synthesis. Yu and coworkers have circumvented this inconvenience by using benzeneseleninic acid **130a** as a catalyst, with good to excellent results even for substrates that were recalcitrant using the **1aa/**H_2O_2 system or that are sensitive to H_2O_2, like aldoximes **184m–o** [135]. In a general way, a variety of aromatic and aliphatic aldoximes **184** were dehydrated to the respective nitriles **185** after 3–24 h of reaction at refluxing acetonitrile in the presence of 5 mol% of **123a**, giving products in 25–98% yield, the lower yields related to substrates unreactive in the presence of **1aa/**H_2O_2 system (Scheme 1.105). The dehydration of aldoximes **184** catalyzed by **123a** follows the same mechanism showed in Scheme 1.106, except that a first step here is the reduction of seleninic

Scheme 1.104 Mechanism of Se-catalyzed dehydration of aldoximes

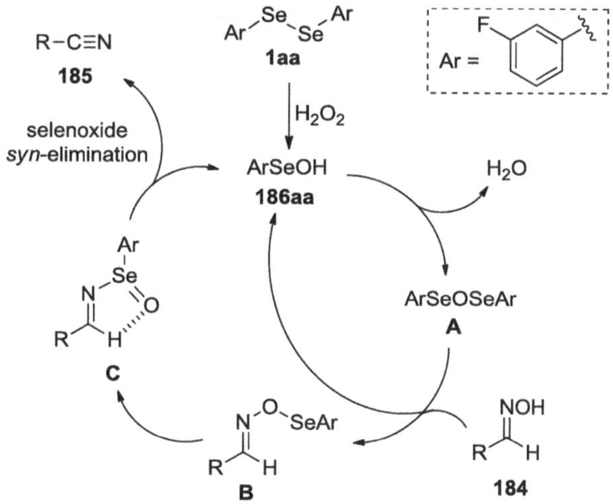

acid by aldoxime **184** to $(PhSe)_2$ **1a**, which is then oxidized by air to selenenic acid **123a**, the actual catalyst of the reaction.

Aldehydes were converted to nitriles in a close related way by Młochowski and coworkers in 1989, by means of the Se-catalyzed oxidation of dimethylhydrazones **187** using 2 mol% of 2-nitrobenzeneseleninic acid (2-NBSeA) **123ab**, derived from bis(2-nitrophenyl)diselenide **1ab**, as the catalyst and 30% H_2O_2 (1.8 equiv.) as the oxidant [136]. Methanolic solutions of eight differently substituted aromatic dimethylhydrazones **187** were efficiently converted to the respective nitriles **185** in 65–97% yields after 30 min to 4 days of reaction at room temperature using the **123ab**/H_2O_2 system. The authors have compared the 2-NBSeA-catalyzed reaction with the uncatalyzed oxidation using overstoichiometric *m*-CPBA and the SeO_2/H_2O_2 system. 2-Nitrobenzeneseleninic acid **123ab** presented the best performance in all the tested examples, except in the case of the aliphatic (*E*)-2-hexylidene-1,1-dimethylhydrazone, that afforded only the parent hexanal as the product (Scheme 1.106).

1.7.2 Se-Catalyzed Redox Dehydration

Very recently, Liebeskind and coworkers have developed a new Se-catalyzed aerobic redox dehydrative synthesis of amides **189** from carboxylic acids **38** and amines **188** [137]. Some advantages of the redox condensation over the classical ones include the reduction in waste and the possibility of performing the reaction under neutral pH. Inspired in a previous work on a benzothiazole/copper co-catalyzed analogue reaction, the authors designed and synthesized a series of properly *o*-substituted electron-deficient diselenides with an amide pendent group, and tested them for their ability in catalyzing the redox dehydration. In the presence of

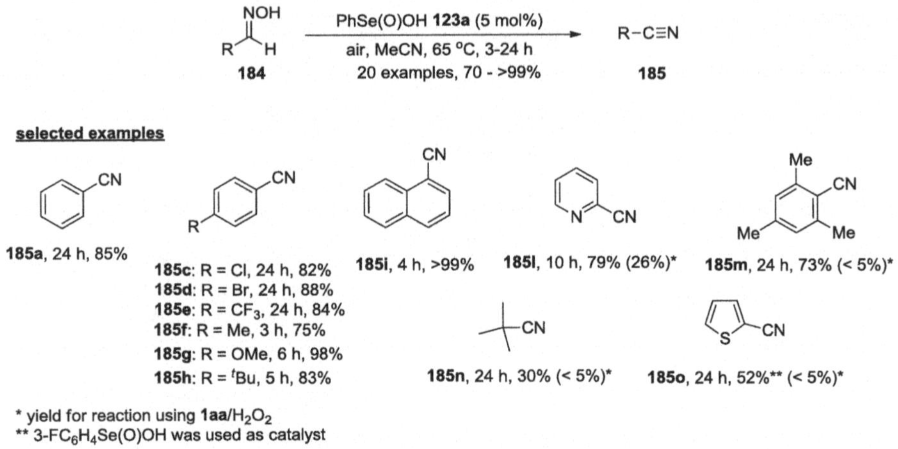

Scheme 1.105 Se-catalyzed dehydration of aldoximes

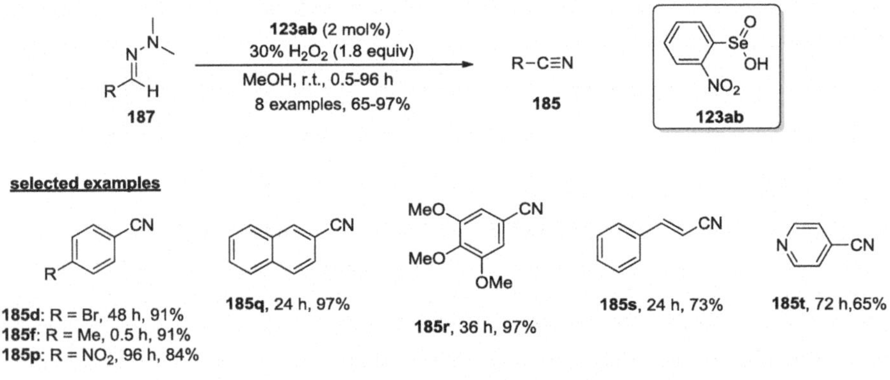

Scheme 1.106 Se-catalyzed oxidation of dimethylhydrazones

diselenide **1ac** or **1ad** (2.5 mol%), $(EtO)_3P$ (1.5 equiv.) as the terminal reductant and dry air (balloon) as the oxidant, a variety of carboxylic acids **38** were reacted with different secondary amines **188** for 16–28 h at 30 or 55 °C to give the respective amides **189** in 63–96% yields. A total of 17 amides and peptides were prepared and it was observed that diselenide **1ac**, with a secondary amide at the *ortho*-position to selenium, was the best catalyst to prepare peptides, while **1ad**, with a tertiary *o*-amide group, was more active in the synthesis of amides (Scheme 1.107).

In the proposed mechanism, the diselenide is firstly reduced by $(EtO)_3P$ in the presence of carboxylic acid **38**, forming the salt **AB**, where the selenide anion **A** is stabilized by the presence of the -NMe_2 group in the pendent *o*-amide group of the diselenide precursor **1ac** or **1ad**. Following, the highly reactive carboxy-triethoxyphosphonium cation **B** reacts with the amine **188** to deliver the desired amide **189** and triethylphosphate. Parallelly, the deprotonated selenol **A** is rapidly oxidized by molecular oxygen from air, regenerating diselenide **1** for a new catalytic cycle (Scheme 1.108).

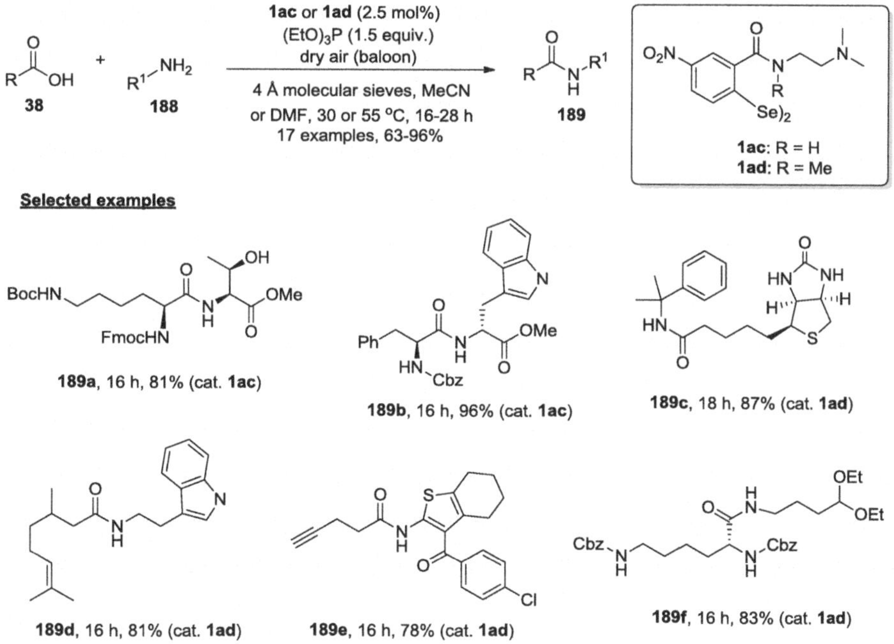

Scheme 1.107 Se-catalyzed aerobic redox dehydrative synthesis of amides

1.7.3 Se-Catalyzed Tishchenko Reaction

Curran and Connon have developed a selenolate-ion-catalyzed Tishchenko reaction, i.e., the base-catalyzed (generally an alkoxide) disproportionation of two aldehydes to generate a coupled ester product [138]. The authors have discovered that dibenzyl diselenide **1o** (2.5 mol%) and bis(3-fluorophenyl)diselenide **1aa** (10 mol%) were the best catalysts for the homo- and the intermolecular crossed-Tishchenko reactions, respectively. The selenolate anions were generated in situ by reaction of **1o** or **1aa** with Bu_2Mg and the reactions were conducted to room temperature in THF. Good results in the homo-coupling were obtained (63–99% yields of **190** after 24 h), even for aldehydes **119** that are recalcitrant under the conventional Tishchenko reaction (coordinating or alkyl ones). Excellent selectivity and yields of **192** were observed in the cross version of the reaction, using non enolizable aromatic aldehydes **119** and trifluoromethyl ketones **191**. The authors have prepared 14 esters **192** by the crossed-Tishchenko reaction, derived from electron-rich and electron-poor benzaldehydes, in 56–96% yields. In both reaction variants, the catalytic efficacy was improved when 3 Å molecular sieves were added to the reaction mixture, due the inhibition of the intermediate acyl selenoate hydrolysis by any water present in the medium (Scheme 1.109).

The reaction mechanism of the Se-catalyzed Tishchenko reaction involves an initial attack of selenolate **A** to aldehyde **119** to form the seleno-hemiacetal anion **B**.

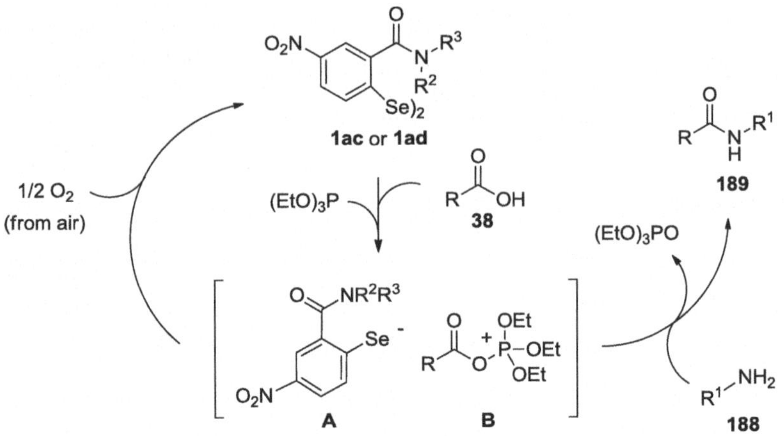

Scheme 1.108 Mechanism of the Se-catalyzed aerobic redox dehydrative synthesis of amides

Following, a hydride transfer from **B** to another molecule of aldehyde **119** (R = H) or to trifluoromethylketone **191** (R = CF_3) occurs, forming alkoxide **C** and selenoester **D**. Coupling of **C** and **D** affords products **190** or **192** and regenerates the selenolate **A** for a new reaction (Scheme 1.110). For the reaction between aldehydes **119** and trifluoromethylketones **191**, dibenzyldiselenide **1o** was not an effective catalyst, because a more electrophilic selenoester **D** is necessary for the crossed reaction.

1.7.4 Se-*Catalyzed Oxidations in Fluorous Biphasic Systems*

Two *F*-tagged butyl-arylselenides **71d** and **71e**, with very similar structures, were prepared by Knochel [139] and Sheldon [140] and used in the activation of H_2O_2 for oxygen-transfer reactions to prepare epoxides, carboxylic acids, and phenols. The use of perfluorinated selenides immobilized in the fluorous phase can be used in fluorous biphasic systems in the Se-catalyzed oxidation of alkenes, aldehydes, and ketones. The fluorous biphasic catalysis was first described by Horváth in 1994 [141] and in such systems the catalyst is immobilized in the fluorous phase (a perfluorinated solvent). After the reaction, the catalyst is easily separated from the reaction mixture and reused in new reactions. An important environmental advantage is the possibility of reusing the *F*-tagged selenides **71d** and **71e** several times after simple thermal separation of the catalyst-containing fluorous phase from the organic, product-containing one.

Knochel and coworkers prepared 2,4-bis(perfluorooctyl)phenyl butylselenide **71d** and used it as a catalyst (5 mol%) in the epoxidation of alkenes **54** with H_2O_2 (60% in water; 1.5–2.0 equiv.) in the presence of a mixture of benzene and bromoperfluoroctane as the solvent [142]. The authors observed that the use of 60% H_2O_2 was needed to avoid the formation of emulsion, observed when 30% H_2O_2 was used.

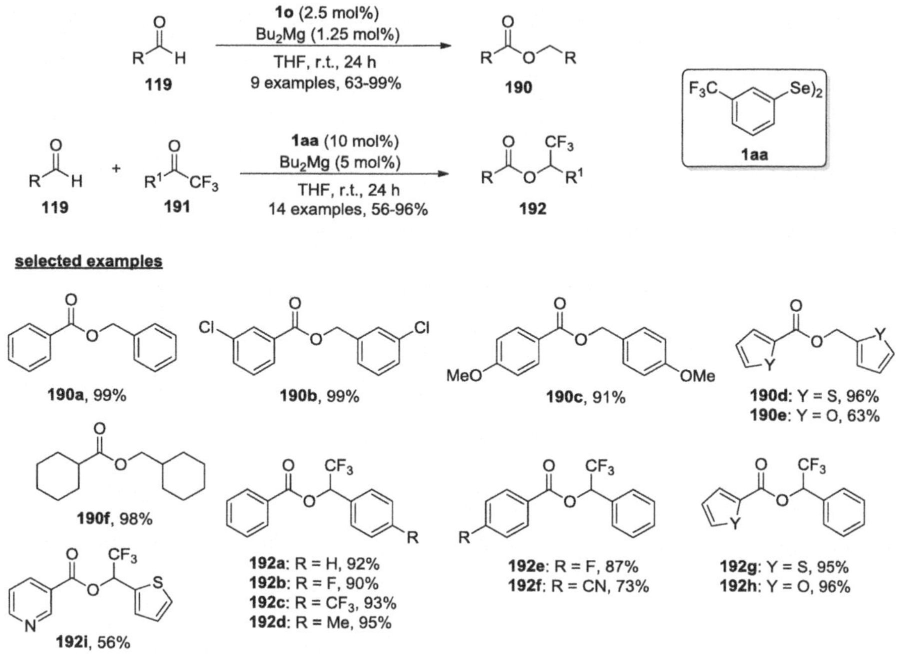

Scheme 1.109 Selenolate ion-catalyzed Tishchenko reaction

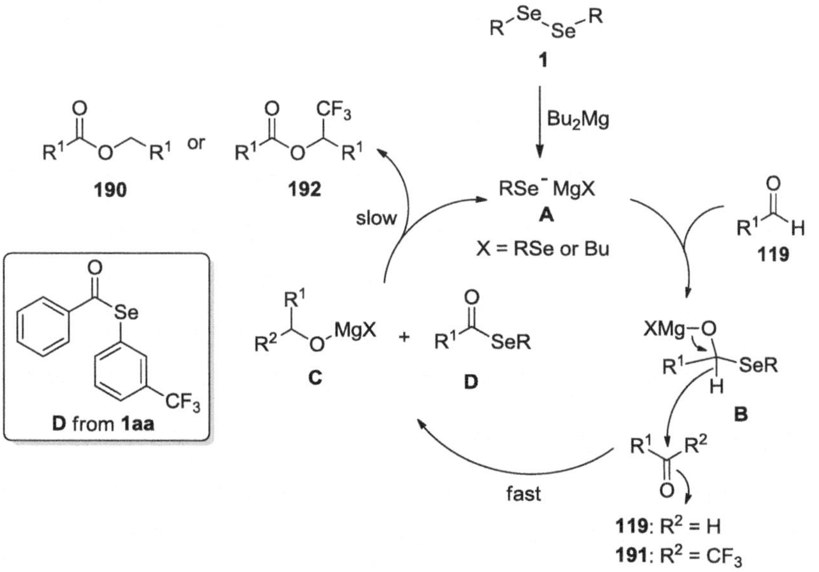

Scheme 1.110 Mechanism of Se-catalyzed Tishchenko reaction

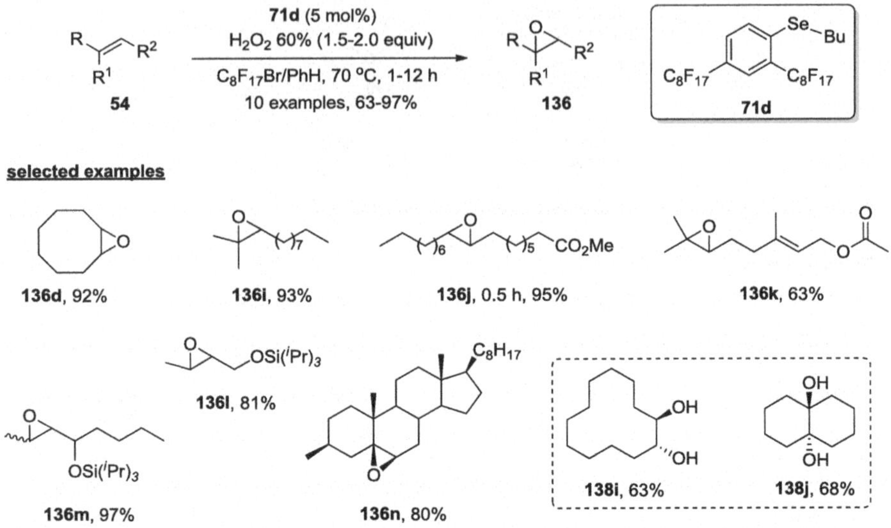

Scheme 1.111 Se-catalyzed epoxidations in fluorous triphasic system

Ten differently substituted alkenes **54** were converted to the respective epoxides **136** in 63–97% yields after 1–12 h of stirring at 70 °C and the method tolerates the presence of esters and silyl enol ethers tethered in the alkene. When cyclododecene and octahydronaphthalene were used, however, the respective *trans* 1,2-diols **138** were obtained instead the expected epoxides (Scheme 1.111). The epoxidation of cyclooctene **54d** to prepare cyclooctene oxide **136d** was repeated ten times using the same catalyst in fluorous solution, giving excellent yields of product (90–93%) in only 1 h of reaction. The first run required 2 h to be complete, due the time for the initial formation of the perfluorinated seleninic acid **123** and then to perseleninic acid **122**, the effective catalyst, as showed in Schemes 1.80 and 1.82.

In the work of Sheldon and coworkers, 2,4-bis(perfluorooctyl)phenyl butylselenide **71e** was prepared and used as a catalyst in the activation of 30% H_2O_2 as an oxidant in the synthesis of carboxylic acids and phenols from aldehydes and in the Baeyer–Villiger reaction of cyclopentenone and cyclohexenone [140]. The authors tested three different fluorous systems: monophasic ($CF_3CHOHCF_3$), biphasic (C_6F_{14}/CF_3CH_2OH), and triphasic ($C_6F_{14}/ClCH_2CH_2Cl$). The triphasic system is the more suitable to carry out the oxidation, even using 30% H_2O_2, once no emulsion was observed under these conditions. The catalyst phase was easily separated and reused for several times in the oxidation of *p*-nitrobenzaldehyde, with slightly decreasing in the activity, probably due to its decomposition. Similar to previously observed by Sheldon in the oxidation catalyzed by bis[3,5-bis(trifluoromethyl) phenyl] diselenide **1s** (Scheme 1.76) [48], electron-rich 3,4,5-trimethoxybenzaldehyde afforded 3,4,5-trimethoxyphenol **135a** instead the carboxylic acid **120**. The same behavior was observed when *p*-anisaldehyde **119d** was the starting material, with *p*-methoxyphenol **135c** being the only isolated product of the reaction

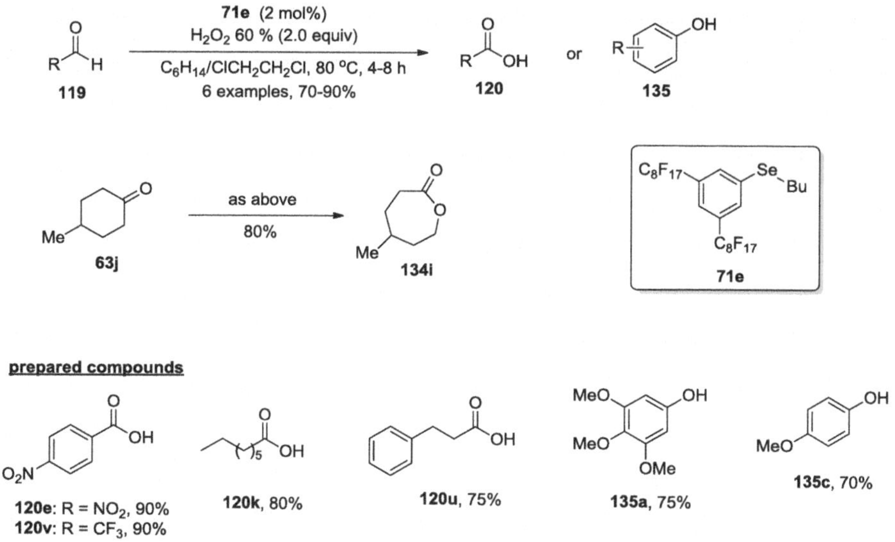

Scheme 1.112 Se-catalyzed oxidations in fluorous triphasic system

(Scheme 1.112). The mechanism of the Se-catalyzed oxidations in fluorous triphasic system is probably the same proposed by the non fluorous reactions, involving the formation of seleninic and perseleninic acids as active species. The highly electron-deficient selenides **71d** and **71e** facilitates the attack by the oxygen of hydrogen peroxide to form seleninic and perseleninic acids.

1.7.5 Se-Catalyzed Oxidative Carbonylation of Anilines

Kim and coworkers have prepared a new anionic selenium-containing ionic liquid (Se-IL) and used it in the oxidative carbonylation of several aromatic amines to prepare urea derivatives in high conversion rates and selectivity [142]. Among the several Se-IL prepared, the best catalytic activity was observed using 2.5 mol% of 1-ethyl-3-methylimidazolium methoxyselenite, [emim][SeO_2(OMe)] **193**, which presented a best performance than 5% Pd/C-KI (the catalyst system for the Asahi Co. process) [143]. The Se-IL **193** was easily prepared by the metathesis reaction of $KSeO_2$(OMe) with [emim][Cl]. Potassium methoxyselenite precursor was quantitatively obtained by the reaction of SeO_2 with K_2CO_3 in methanol. Five aniline derivatives **188** were reacted under the oxidative carbonylation conditions ($P = 1.36$ MPa; O_2–CO 20:80 v/v) in the presence of **193** (4 mol%) and methanol for 2 h to afford the respective symmetrical ureas **194** in conversion rates of 28–97% and 99–100% of selectivity (Scheme 1.113). The original catalytic activity of **193** remained even after five successive reactions, indicating a good recyclability of the Se-IL simply by filtering off the solid urea formed after each reaction.

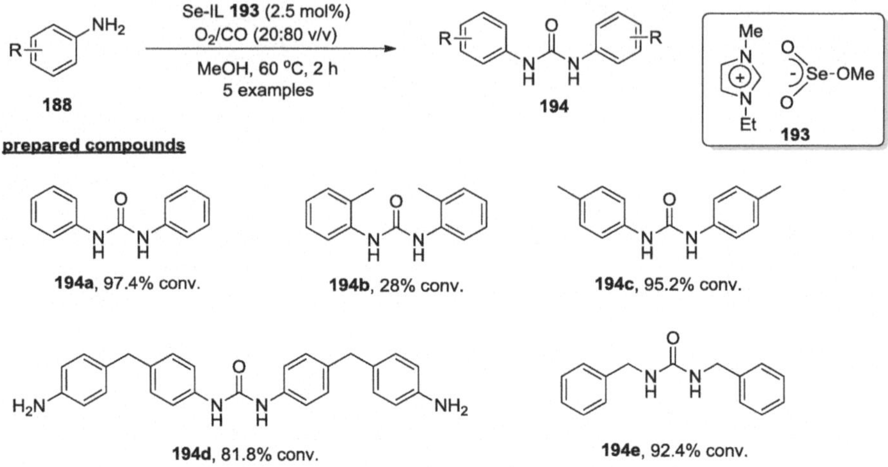

Scheme 1.113 Oxidative carbonylation in Se-containing ionic liquids

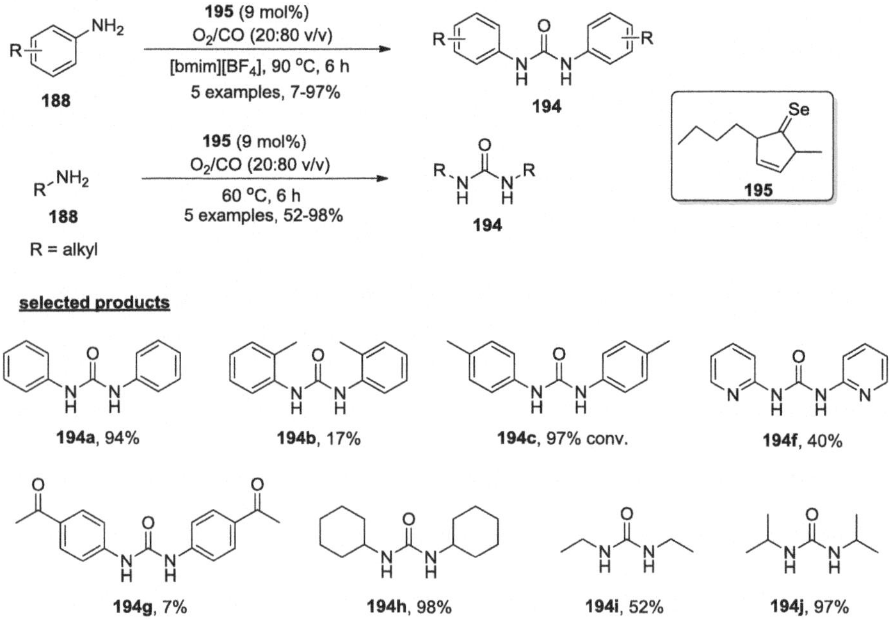

Scheme 1.114 *Se*-catalyzed oxidative carbonylation of aromatic anilines

Recently, Tian and coworkers developed a green Se-catalyzed oxidative carbonylation of aromatic anilines **188** to the respective symmetrical ureas **194** using ionic liquid (IL) as the solvent and air/CO [144]. The authors have prepared 1-butyl-3-methylimidazole-2-selenone **195** and used it in a solution of [bmim][BF_4] in the reaction of six aromatic amines and benzylamine at 90 °C for 6 h. The Se-catalyst **195**

(9 mol%) was effective in promoting the reaction of 2-amino-pyridine and an aqueous solution of ethylamine. When aliphatic amines were used as the substrate, the reactions occurred satisfactorily at 60 °C, even in the absence of IL, except for piperidine, which was unreactive under these conditions (Scheme 1.114). The IL-catalyst solution was reused in additional reactions of aniline **188a** after separation of 1,3-diphenylurea **194a** by addition of water and acetone to the reaction mixture. The authors observed a slightly increased in the yields of **194a** in the successive reactions (94, 96, 99, and 92%), which was attributed to the participation of unreacted substrate in the subsequent reactions.

References

1. Gromer S, Johansson L, Bauer H, Arscott LD, Rauch S, Ballou DP, Williams CH Jr, Schirmer RH, Arner ES (2003) Active sites of thioredoxin reductases: why selenoproteins? Proc Natl Acad Sci U S A 100:12618–12623
2. Hondal RJ, Marino SM, Gladyshev VN (2013) Selenocysteine in thiol/disulfide-like exchange reactions. Antiox Redox Signal 18:1675–1689
3. Jacob C, Giles GI, Giles NM, Sies H (2003) Sulfur and selenium: the role of oxidation state in protein structure and function. Angew Chem Int Ed Engl 42:4742–4758
4. Santi C, Tidei C, Scalera C (2013) Selenium containing compounds: from poison to drug candidates (a review on the GPx-like activity). Curr Chem Biol 7:25–36
5. Müller A, Cadenas E, Graf P, Sies H (1984) A novel biologically active seleno-organic compound-I. Glutathione peroxidase-like activity in vitro and antioxidant capacity of PZ 51 (Ebselen). Biochem Pharmacol 33:3235–3239
6. Wendel A, Fausel M, Safayhi H, Tiegs G, Otter R (1984) A novel biologically active seleno-organic compound-II. Activity of PZ 51 in relation to glutathione peroxidase. Biochem Pharmacol 33:3241–3245
7. Bhowmick D, Mugesh G (2015) Insights into the catalytic mechanism of synthetic glutathione peroxidase mimetics. Org Biomol Chem 13:10262–10272
8. Santi C, Santoro S, Battiste B (2010) Organoselenium compounds as catalysts in nature and laboratory. Curr Org Chem 14:2442–2462
9. Nishibayashi Y, Uemura S (2000) Selenium compounds as ligands and catalysis. In: Wirth T (ed) Organoselenium chemistry - modern developments in organic synthesis, Topics in current chemistry, vol 208. Springer, Heidelberg, pp 235–255
10. With T (1999) Chiral selenium compounds in organic synthesis. Tetrahedron 55:1–28
11. Singh FV, Wirth T (2012) Selenium compounds as ligands and catalysts. In: Wirth T (ed) Organoselenium chemistry - synthesis and reactions. Wiley-VCH Verlag & Co., Weinheim, pp 321–360
12. Santoro S, Azeredo JB, Nascimento V, Sancineto L, Braga AL, Santi C (2014) The green side of the moon: ecofriendly aspects of organoselenium chemistry. RSC Adv 4:31521–31535
13. Młochowski J, Wójtowicz-Młochowska H (2015) Developments in synthetic application of selenium(IV) oxide and organoselenium compounds as oxygen donors and oxygen-transfer agents. Molecules 20:10205–10243
14. Młochowski J, Brząszcz M, Chojnacka M, Giurg M, Wójtowicz H (2004) Diaryl diselenides and benzisoselenazol-3(2H)-ones as oxygen transfer agents. ARKIVOC 2004:226–248
15. Młochowski J, Brząszcz M, Giurg M, Palus J, Wójtowicz H (2008) Selenium-promoted oxidation of organic compounds: reactions and mechanisms. Eur J Org Chem 2003:4329–4339
16. Giurg M, Syper L (2008) Diaryl diselenides and related compounds as oxygen-transfer agents. Phosphorus Sulfur Silicon Relat Elem 183:970–985

17. Młochowski J, Peczynska-Czoch W, Pietka-Ottlik M, Wójtowicz-Młochowska H (2011) Non-metal and enzymatic catalysts for hydroperoxide oxidation of organic compounds. Open Cat J 4:54–82
18. Freudendahl DM, Santoro S, Shahzad SA, Santi S, Wirth T (2009) Green chemistry with selenium reagents: development of efficient catalytic reactions. Angew Chem Int Ed 48:8409–8411
19. Back TG (2009) Design and synthesis of some biologically interesting natural and unnatural products based on organosulfur and selenium chemistry. Can J Chem 87:1657–1674
20. Alberto EE, Braga AL (2011) Activation of peroxides by organoselenium catalysts: a synthetic and biological perspective. In: Woollins JD, Laitinen RS (eds) Selenium and tellurium chemistry. Springer, Berlin, pp 251–283
21. Tiecco M (2000) Electrophilic selenium, selenocyclizations. In: Wirth T (ed) Organoselenium chemistry - modern developments in organic synthesis, Topics in current chemistry, vol 208. Springer, Heidelberg, pp 7–53
22. Santi C, Tidei C (2014) Electrophilic selenium/tellurium reagents: reactivity and their contribution to green chemistry. In: Rappoport Z, Liebman JF, Marek I, Patai S (eds) The chemistry of organic selenium and tellurium compounds, vol 4. Wiley, New York, pp 569–656
23. Guo R, Liao L, Zhao X (2017) Electrophilic selenium catalysis with electrophilic N-F reagents as the oxidants. Molecules 22:835. https://doi.org/10.3390/molecules22050835
24. Ortgies S, Breder A (2017) Oxidative alkene functionalizations via selenium-π-acid catalysis. ACS Catal 7:5828–5840
25. Freudendahl DM, Shahzad SA, Wirth T (2009) Recent advances in organoselenium chemistry. Eur J Org Chem 2009:1649–1664
26. Breder A, Ortgies S (2015) Recent developments in sulfur- and selenium-catalyzed oxidative and isohypsic functionalization reactions of alkenes. Tetrahedron Lett 56:2843–2852
27. Tidei C, Santi C (2014) Selenium and "bio-logic" catalysis: new bioinspired catalytic reactions. In: Santi C (ed) Organoselenium chemistry: between synthesis and biochemistry. Bentham Science, Sharjah, pp 345–360. https://doi.org/10.2174/97816080583891140101
28. Denmark SE, Jaunet A (2014) Catalytic, enantioselective, intramolecular carbosulfenylation of olefins. Preparative and stereochemical aspects. J Org Chem 79:140–171
29. Chabaud B, Sharpless KB (1979) Selenium-catalyzed nonradical chlorination of olefins with N-chlorosuccinimide. J Org Chem 44:4204–4208
30. Hori T, Sharpless KB (1979) Conversion of allylic phenylselenides to the rearranged allylic chlorides by N-chlorosuccinimide. Mechanism of selenium-catalyzed allylic chlorination of α-pinene. J Org Chem 44:4208–4210
31. Tiecco M, Testaferri L, Temperini A, Marini F, Bagnoli L, Santi C (1999) Selenium promoted stereospecific one-pot conversion of cinnamyl derivatives into oxazoles. A simple synthetic route to racemic taxol side chain. Synth Commun 29:1773–1778
32. Tiecco M, Testaferri L, Santi C (1999) Catalytic oxyselenenylation-deselenenylation reactions of alkenes-stereoselective one-pot conversion of 3-alkenols into 2,5-dihydrofurans. Eur J Org Chem 1999:797–803
33. Tiecco M, Testaferri L, Tingoli M, Marini F (1994) Selenium promoted conversion of α-substituted β,γ-unsaturated ketones into 2,3,5-trisubstituted furans. Synlett 1994:373–374
34. Tiecco M, Testaferri L, Tingoli M, Bagnoli L, Santi C (1993) Catalytic conversion of β,γ-unsaturated esters, amides and nitriles into γ-alkoxy or γ-hydroxy α,β-unsaturated derivatives induced by persulfate anion oxidation of diphenyl diselenide. J Chem Soc Chem Commun 1993:637–639
35. Iwaoka M, Tomoda S (1992) Catalytic conversion of alkenes into allylic ethers and esters using diselenides having internal tertiary amines. J Chem Soc Chem Commun 1992:1165–1167
36. Tiecco M, Testaferri L, Tingoli M, Bartoli D, Marini F (1991) Selenium-promoted conversion of β-diketones and β-keto esters into α,α-dimethoxy β-diketones and α,α-dimethoxy β-keto esters. J Org Chem 56:5207–5210

37. Tiecco M, Testaferri L, Tingoli M, Chianelli D, Bartoli D (1991) Selenium-mediated conversion of alkynes into α-dicarbonyl compounds. J Org Chem 56:4529–4534
38. Tiecco M, Testaferri L, Tingoli M, Bartoli D (1990) Selenium-catalyzed conversion of methyl ketones into α-keto acetals. J Org Chem 55:4523–4528
39. Tingoli M, Mazzella M, Panunzi B, Tuzi A (2011) Elemental iodine or diphenyl diselenide in the [bis(trifluoroacetoxy)-iodo] benzene-mediated conversion of alkynes into 1,2-diketones. Eur J Org Chem 2011:399–404
40. Singh FV, Wirth T (2011) Selenium-catalyzed regioselective cyclization of unsaturated carboxylic acids using hypervalent iodine oxidants. Org Lett 13:6504–6507
41. Shahzad SA, Venin C, Wirth T (2010) Diselenide- and disulfide-mediated synthesis of isocoumarins. Eur J Org Chem 2010:3465–3472
42. Browne DM, Niyomura O, Wirth T (2007) Catalytic use of selenium electrophiles in cyclizations. Org Lett 9:3169–3171
43. Yu L, Li H, Zhang X, Ye J, Liu J, Xu Q, Lautens M (2014) Organoselenium-catalyzed mild dehydration of aldoximes: an unexpected practical method for organonitrile synthesis. Org Lett 16:1346–1349
44. Yu L, Wu Y, Cao H, Zhang X, Shi X, Luan J, Chen T, Pan Y, Xu Q (2014) Facile synthesis of 2-methylenecyclobutanones via $Ca(OH)_2$-catalyzed direct condensation of cyclobutanone with aldehydes and $(PhSe)_2$-catalyzed Baeyer-Villiger oxidation to 4-methylenebutanolides. Green Chem 16:287–293
45. Santi C, Lorenzo RD, Tidei C, Bagnoli L, Wirth T (2012) Stereoselective selenium catalyzed dihydroxylation and hydroxymethoxylation of alkenes. Tetrahedron 68:10530–10535
46. Santoro S, Santi C, Sabatini M, Testaferri L, Tiecco M (2008) Eco-friendly olefin dihydroxylation catalyzed by diphenyl diselenide. Adv Synth Catal 350:2881–2884
47. Ichikawa H, Usami Y, Arimoto M (2005) Synthesis of novel organoselenium as catalyst for Baeyer-Villiger oxidation with 30% H_2O_2. Tetrahedron Lett 46:8665–8668
48. ten Brink GJ, Vis JM, Arends IWCE, Sheldon RA (2001) Selenium-catalyzed oxidations with aqueous hydrogen peroxide. 2. Baeyer-Villiger reactions in homogeneous solution. J Org Chem 66:2429–2433
49. Torii S, Uneyama K, Ono M, Bannou T (1981) Generation and recycle use of selenenylating reagents in electrochemical oxyselenenylation-deselenenylation of olefins. J Am Chem Soc 103:4606–4608
50. Van der Toorn JC, Kemperman G, Sheldon RA, Arends IWCE (1990) Electroreductive ring-opening of α,β-epoxy carbonyl compounds and their homologues through recyclable use of diphenyl diselenide or diphenyl ditelluride as a mediator. J Org Chem 55:1548–1553
51. Niyomura O, Cox M, Wirth T (2006) Electrochemical generation and catalytic use of selenium electrophiles. Synlett 2006:251–254
52. Van der Toorn JC, Kemperman G, Sheldon RA, Arends IWCE (2009) Diphenyldiselenide-catalyzed selective oxidation of activated alcohols with tert-butyl hydroperoxide: new mechanistic insights. J Org Chem 74:3085–3089
53. Torii S, Uneyama K, Ono M (1980) Novel synthesis of DL-marmelolactone and DL-rose oxide by electrochemical oxyselenenylation-deselenenylation sequence. Tetrahedron Lett 21:2653–2654
54. Fujita K, Iwaoka M, Tomoda S (1994) Synthesis of diaryl diselenides having chiral pyrrolidine rings with C_2 symmetry. Their application to the asymmetrical methoxyselenenylation of *trans*-β-methylstyrenes. Chem Lett 1994:923–926
55. Fukuzawa S, Takahashi K, Kato H, Yamazaki H (1997) Asymmetric methoxyselenenylation of alkenes with chiral ferrocenylselenium reagents. J Org Chem 62:7711–7716
56. Wirth T, Häuptli S, Leuenberger M (1998) Catalytic asymmetric oxyselenenylation-elimination reactions using chiral selenium compounds. Tetrahedron-Asymmetry 9:547–550
57. Tiecco M, Testaferri L, Santi C, Tomassini C, Marini F, Bagnoli L, Temperini A (2000) New nitrogen containing chiral diselenides: synthesis and asymmetric addition reactions to olefins. Tetrahedron-Asymmetry 11:4645–4650

58. Tiecco M, Testaferri L, Santi C, Tomassini C, Marini F, Bagnoli L, Temperini A (2002) Preparation of a new chiral non-racemic sulfur-containing diselenide and applications in asymmetric synthesis. Chem Eur J 8:1118–1124
59. Tunge JA, Mellegaard SR (2004) Selective selenocatalytic allylic chlorination. Org Lett 6:1205–1207
60. Trenner J, Depken C, Weber T, Breder A (2013) Direct oxidative allylic and vinylic amination of alkenes through selenium catalysis. Angew Chem Int Ed 52:8952–8956
61. Deng Z, Wei J, Liao L, Huang H, Zhao X (2015) Organoselenium-catalyzed, hydroxy-controlled regio- and stereoselective amination of terminal alkenes: efficient synthesis of 3-amino allylic alcohols. Org Lett 17:1834–1837
62. Ortgies S, Breder A (2015) Selenium-catalyzed oxidative C(sp^2)-H amination of alkenes exemplified in the expedient synthesis of (aza-)indoles. Org Lett 17:2748–2751
63. Zhang X, Guo R, Zhao X (2015) Organoselenium-catalyzed synthesis of indoles through intramolecular C-H amination. Org Chem Front 2:1334–1337
64. Guo R, Huang J, Huang H, Zhao X (2016) Organoselenium-catalyzed synthesis of oxygen- and nitrogen-containing heterocycles. Org Lett 18:504–507
65. Liao L, Guo R, Zhao X (2017) Organoselenium-catalyzed regioselective C-H pyridination of 1,3-dienes and alkenes. Angew Chem Int Ed 56:3201–3205
66. Krätzschmar F, Kaßel M, Delony D, Breder A (2015) Selenium-catalyzed C(sp^3)-H acyloxylation: application in the expedient synthesis of isobenzofuranones. Chem Eur J 21:7030–7035
67. Kawamata Y, Hashimoto T, Maruoka K (2016) A chiral electrophilic selenium catalyst for highly enantioselective oxidative cyclization. J Am Chem Soc 138:5206–5209
68. Ortgies S, Depken C, Breder A (2016) Oxidative allylic esterification of alkenes by cooperative selenium-catalysis using air as the sole oxidant. Org Lett 18:2856–2859
69. Leisering S, Riaño I, Depken C, Gross LJ, Weber M, Lentz D, Zimmer R, Stark CBW, Breder A, Christmann M (2017) Synthesis of (+)-Greek tobacco lactone via a diastereoablative epoxidation and a selenium-catalyzed oxidative cyclization. Org Lett 19:1478–1481
70. Browne DM, Niyomura O, Wirth T (2008) Catalytic addition-elimination reactions towards butenolides. Phosphorus Sulfur Silicon 183:1026–1035
71. Fragale G, Wirth T (1998) Chiral diselenides in asymmetric cyclization reactions. Eur J Org Chem 1998:1361–1369
72. Cresswell AJ, Eey STC, Denmark SE (2015) Catalytic, stereospecific *syn*-dichlorination of alkenes. Nat Chem 7:146–152
73. Wang C, Tunge J (2004) Selenocatalytic α-halogenation. Chem Commun 2004:2694–2695
74. Mellegaard-Waetzig SR, Wang C, Tunge JA (2006) Selenium-catalyzed oxidative halogenation. Tetrahedron 62:7191–7198
75. Barrero AF, Quílez del Moral JF, Mar Herrador M, Cortés M, Arteaga P, Catalán JV, Sánchez EM, Arteaga JF (2006) Solid-phase selenium-catalyzed selective allylic chlorination of polyprenoids: facile syntheses of biologically active terpenoids. J Org Chem 71:5811–5814
76. Denmark SE, Beutner GL (2008) Lewis base catalysis in organic synthesis. Angew Chem Int Ed 47:1560–1638
77. Mellegaard SR, Tunge JA (2004) Selenium-catalyzed halolactonization: nucleophilic activation of electrophilic halogenating reagents. J Org Chem 69:8979–8981
78. Carrera I, Brovetto MC, Seoane GA (2006) Selenium-catalyzed iodohydrin formation from alkenes. Tetrahedron Lett 47:7849–7852
79. Tay DW, Tsoi IT, Er JC, Leung GYC, Yeung YY (2013) Lewis basic selenium catalyzed chloroamidation of olefins using nitriles as the nucleophiles. Org Lett 15:1310–1313
80. Luo J, Zhu Z, Liu Y, Zhao X (2015) Diaryl selenide catalyzed vicinal trifluoromethylthioamination of alkenes. Org Lett 17:3620–3623
81. Zhu Z, Luo J, Zhao X (2017) Combination of Lewis basic selenium catalysis and redox selenium chemistry: synthesis of trifluoromethylthiolated tertiary alcohols with alkenes. Org Lett 19:4940–4943

82. Balkrishna SJ, Prasad CD, Panini P, Detty MR, Chopra D, Kumar S (2012) Isoselenazolones as catalysts for the activation of bromine: bromolactonization of alkenoic acids and oxidation of alcohols. J Org Chem 77:9541–9552
83. Boualy B, El Houssame S, Sancineto L, Santi C, Ait Ali M, Stoeckli-Evans H, El Firdoussi L (2016) A mild and efficient method for the synthesis of a new optically active diallyl selenide and its catalytic activity in the allylic chlorination of natural terpenes. New J Chem 40:3395–3399
84. Denmark SE, Collins WR (2007) Lewis base activation of Lewis acids: development of a Lewis base catalyzed selenolactonization. Org Lett 9:3801–3804
85. Denmark SE, Kornfilt DJP, Vogler T (2011) Catalytic asymmetric thiofunctionalization of unactivated alkenes. J Am Chem Soc 133:15308–15311
86. Denmark SE, Jaunet A (2013) Catalytic, enantioselective, intramolecular carbosulfenylation of olefins. J Am Chem Soc 135:6419–6422
87. Denmark SE, Kornfilt DJP (2017) Catalytic, enantioselective, intramolecular sulfenofunctionalization of alkenes with phenols. J Org Chem 82:3192–3222
88. Denmark SE, Chi HM (2014) Lewis base catalyzed, enantioselective, intramolecular sulfenoamination of olefins. J Am Chem Soc 136:8915–8918
89. Denmark SE, Chi HM (2017) Catalytic, enantioselective, intramolecular sulfenoamination of alkenes with anilines. J Org Chem 82:3826–3843
90. Denmark SE, Chi HM (2014) Catalytic, enantioselective, intramolecular carbosulfenylation of olefins. Mechanistic aspects: a remarkable case of negative catalysis. J Am Chem Soc 136:3655–3663
91. Denmark SE, Hartmann E, Kornfilt DJP, Wang H (2014) Mechanistic, crystallographic, and computational studies on the catalytic, enantioselective sulfenofunctionalization of alkenes. Nat Chem 6:1056–1064
92. Denmark SE, Rossi S, Webster MP, Wang H (2014) Catalytic, enantioselective sulfenylation of ketone-derived enoxysilanes. J Am Chem Soc 136:13016–13028
93. Chen F, Tan CK, Yeung YY (2013) C_2-Symmetric cyclic selenium-catalyzed enantioselective bromoaminocyclization. J Am Chem Soc 135:1232–1235
94. Luo J, Liu Y, Zhao X (2017) Chiral selenide-catalyzed enantioselective construction of saturated trifluoromethylthiolated azaheterocycles. Org Lett 19:3434–3437
95. Goti A, Cardona F (2008) Hydrogen peroxide in green oxidation reactions: recent catalytic processes. In: Tundo P, Esposito V (eds) Green chemical reactions, NATO science for peace and security series (Series C: Environmental security). Springer, Dordrecht, pp 191–212
96. Venturello C, Gambaro M (1991) Selective oxidation of alcohols and aldehydes with hydrogen peroxide catalyzed by methyltrioctylammonium tetrakis(oxodiperoxotungsto)-phosphate(3-) under two-phase conditions. J Org Chem 56:5924–5931
97. Che CM, Yip WP, Yu WY (2006) Ruthenium-catalyzed oxidation of alkenes, alkynes, and alcohols to organic acids with aqueous hydrogen peroxide. Chem Asian J 1:453–458
98. Gopinath R, Patel BK (2000) A catalytic oxidative esterification of aldehydes using V_2O_5–H_2O_2. Org Lett 2:577–579
99. Trost BM, Masuyama Y (1984) Chemoselectivity in molybdenum catalyzed alcohol and aldehyde oxidations. Tetrahedron Lett 25:173–176
100. Giurg M, Młochowski J, Ambrożak A (2002) Hydrogen peroxide oxidation of N,N-dimethylhydrazone promoted by selenium compounds, titanosilicalites or acetonitrile. Pol J Chem 76:1713–1720
101. Dodd RH, Le Hyaric M (1993) The oxidation of aromatic aldehydes to carboxylic acids using hydrogen peroxide in formic acid. Synthesis 1993:295–297
102. Dalcanale E, Montanari F (1986) Selective oxidation of aldehydes to carboxylic acids with sodium chlorite-hydrogen peroxide. J Org Chem 51:567–569
103. Sharpless KB, Lauer RF, Teranishi AY (1973) Electrophilic and nucleophilic organoselenium reagents. New routes to α,β-unsaturated carbonyl compounds. J Am Chem Soc 95:6137–6139

104. Sancineto L, Tidei C, Bagnoli L, Marini F, Lenardão EJ, Santi C (2015) Selenium catalyzed oxidation of aldehydes: green synthesis of carboxylic acids and esters. Molecules 20:10496–10510
105. Taylor RT, Flood LA (1983) Polystyrene-bound phenylseleninic acid: catalytic oxidations of olefins, ketones, and aromatics. J Org Chem 48:5160–5164
106. Wójtowicz H, Soroko G, Młochowski J (2008) New recoverable organoselenium catalyst for hydroperoxide oxidation of organic substrates. Synth Commun 38:2000–2010
107. Rangraz Y, Nemati F, Elhampour A (2018) Diphenyl diselenide immobilized on magnetic nanoparticles: a novel and retrievable heterogeneous catalyst in the oxidation of aldehydes under mild and green conditions. J Colloid Interface Sci 509:485–494
108. Wang T, Jing X, Chen C, Yu L (2017) Organoselenium-catalyzed oxidative C=C bond cleavage: a relatively green oxidation of alkenes into carbonyl compounds with hydrogen peroxide. J Org Chem 82:9342–9349
109. Syper L (1989) The Baeyer-Villiger oxidation of aromatic aldehydes and ketones with hydrogen peroxide catalyzed by selenium compounds. A convenient method for the preparation of phenols. Synthesis 1989:167–172
110. Yu L, Ye J, Zhang X, Dinga Y, Xu Q (2015) Recyclable $(PhSe)_2$-catalyzed selective oxidation of isatin by H_2O_2: a practical and waste-free access to isatoic anhydride under mild and neutral conditions. Cat Sci Technol 5:4830–4838
111. Zhang X, Ye J, Yu L, Shi X, Zhang M, Xu Q, Lautens M (2015) Organoselenium-catalyzed Baeyer-Villiger oxidation of α,β-unsaturated ketones by hydrogen peroxide to access vinyl esters. Adv Synth Catal 357:955–960
112. Reich HJ, Chow F, Peake SL (1978) Seleninic acids as catalysts of olefins and sulfides using hydrogen peroxide. Synthesis 1978:299–301
113. Grieco PA, Yokoyama Y, Gilman S, Nishizawa M (1977) Organoselenium chemistry. Epoxidation of olefins with benzeneseleninic acid and hydrogen peroxide ("benzeneperoxyseleninic acid"). J Org Chem 42:2034–2036
114. Kametani T, Nemoto H, Fukumoto K (1977) A new method for an epoxidation of olefins and its application to a biomimetic type synthesis of monoterpenes, linalyloxides. Heterocycles 6:1365–1370
115. Hori T, Sharpless KB (1978) Synthetic applications of arylselenenic and arylseleninic acids. Conversion of olefins to allylic alcohols and epoxides. J Org Chem 43:1689–1697
116. ten Brink GJ, Fernandes BCM, van Vliet MCA, Arends IWCE, Sheldon RA (2001) Selenium catalysed oxidations with aqueous hydrogen peroxide. Part I: Epoxidation reactions in homogeneous solution. J Chem Soc Perkin Trans 2001:224–228
117. Yu L, Bai Z, Zhang X, Zhang X, Ding Y, Xu Q (2016) Organoselenium-catalyzed selectivity-switchable oxidation of β-ionone. Cat Sci Technol 6:1804–1809
118. García-Marín H, van der Toorn JC, Mayoral JA, García JI, Arends IWCE (2009) Glycerol-based solvents as green reaction media in epoxidations with hydrogen peroxide catalysed by bis[3,5-bis(trifluoromethyl)-diphenyl] diselenide. Green Chem 11:1605–1609
119. Sancineto L, Mangiavacchi F, Tidei C, Bagnoli L, Marini F, Gioiello A, Scianowski J, Santi C (2017) Selenium-catalyzed oxacyclization of alkenoic acids and alkenols. Asian J Org Chem 6:988–992
120. Cerra B, Mangiavacchi F, Santi C, Lozza AM, Gioiello A (2017) Selective continuous flow synthesis of hydroxy lactones from alkenoic acids. React Chem Eng 2:467–471
121. Francavilla C, Bright FV, Detty MR (1999) Dendrimeric catalysts for the activation of hydrogen peroxide. Increasing activity per catalytic phenylseleno group in successive generations. Org Lett 1:1043–1046
122. Francavilla C, Drake MD, Bright FV, Detty MR (2001) Dendrimeric organochalcogen catalysts for the activation of hydrogen peroxide: improved catalytic activity through statistical effects and cooperativity in successive generations. J Am Chem Soc 123:57–67
123. Drake MD, Bright FV, Detty MR (2003) Dendrimeric organochalcogen catalysts for the activation of hydrogen peroxide: origins of the "dendrimer effect" with catalysts terminating in phenylseleno groups. J Am Chem Soc 125:12558–12566

124. Newkome GR, Moorefield C, Vögtle F (2001) Dendritic macromolecules: concepts, synthesis, perspectives. Wiley-VCH, Weinheim
125. Drake MD, Bateman MA, Detty MR (2003) Substituent effects in arylseleninic acid-catalyzed bromination of organic substrates with sodium bromide and hydrogen peroxide. Organometallics 22:4158–4162
126. Bennett SM, Tang Y, McMaster D, Bright FV, Detty MR (2008) A xerogel-sequestered selenoxide catalyst for brominations with hydrogen peroxide and sodium bromide in an aqueous environment. J Org Chem 73:6849–6852
127. Gatley CM, Muller LM, Lang MA, Alberto EE, Detty MR (2015) Xerogel-sequestered silanated organochalcogenide catalysts for bromination with hydrogen peroxide and sodium bromide. Molecules 20:9616–9639
128. Alberto EE, Braga AL, Detty MR (2012) Imidazolium-containing diselenides for catalytic oxidations with hydrogen peroxide and sodium bromide in aqueous solutions. Tetrahedron 68:10476–10481
129. Ribaudo G, Bellanda M, Menegazzo I, Wolters LP, Bortoli M, Ferrer-Sueta G, Zagotto G, Orian L (2017) Mechanistic insight into the oxidation of organic phenylselenides by H_2O_2. Chem Eur J 23:2405–2422
130. Crich D, Sannigrahi M (2002) Rapid assembly of tetrahydrodibenzofurans and tetrahydrocarbazoles from benzene and *o*-iodophenols and *o*-iodoanilines: reductive radical arylation of benzene in action. Tetrahedron 58:3319–3322
131. Crich D, Rumthao S (2004) Synthesis of carbazomycin B by radical arylation of benzene. Tetrahedron 60:1513–1516
132. Crich D, Grant D (2005) Synthesis of a 4,6-disubstituted dibenzofuran β-sheet initiator by reductive radical arylation of benzene. J Org Chem 70:2384–2386
133. Crich D, Patel M (2005) Facile dearomatizing radical arylation of furan and thiophene. Org Lett 7:3625–3628
134. Clive DLJ, Pham MP, Subedi R (2007) Carbocyclization by radical closure onto *O*-trityl oximes: dramatic effect of diphenyl diselenide. J Am Chem Soc 129:2713–2717
135. Zhang X, Sun J, Ding Y, Yu L (2015) Dehydration of aldoximes using PhSe(O)OH as the pre-catalyst in air. Org Lett 17:5840–5842
136. Said SB, Skarzevski J, Młochowski J (1989) Conversion of aldehydes into nitriles via oxidation of their dimethylhydrazones. Synthesis 1989:223–224
137. Akondi SM, Gangireddy P, Pickel TC, Liebeskind LS (2018) Aerobic, diselenide catalyzed redox dehydration: amides and peptides. Org Lett 20:538–541
138. Curran SP, Connon SJ (2012) Selenide ions as catalysts for homo- and crossed-Tishchenko reactions of expanded scope. Org Lett 14:1074–1077
139. Betzemeier B, Lhermitte F, Knochel P (1999) A selenium catalyzed epoxidation in perfluorinated solvents with hydrogen peroxide. Synlett 1999:489–491
140. ten Brink GJ, Vis JM, Arends IWCE, Sheldon RA (2002) Selenium catalyzed oxidations with aqueous hydrogen peroxide. Part 3: Oxidation of carbonyl compounds under mono/bi/triphasic conditions. Tetrahedron 58:3977–3983
141. Horvárth IT, Rábai J (1994) Facile catalyst separation without water: fluorous biphase hydroformylation of olefins. Science 266:72–75
142. Kim HS, Kim YJ, Lee H, Park KY, Lee C, Chin CS (2002) Ionic liquids containing anionic selenium species: applications for the oxidative carbonylation of aniline. Angew Chem Int Ed 41:4300–4303
143. Fukuoka S, Chono M, Kohno M (1984) A novel catalytic synthesis of carbamates by oxydative alkoxycarbonylation of amines in the presence of palladium and iodide. J Chem Soc Chem Commun 1984:399–400
144. Tian F, Chen Y, Wang X, Li P, Lu S (2015) Oxidative carbonylation of aromatic amines with CO catalyzed by 1,3-dialkylimidazole-2-selenone in ionic liquids. J Chem. https://doi.org/10.1155/2015/210806

Chapter 2
Bioactive Organoselenium Compounds and Therapeutic Perspectives

Abstract After 1984, when ebselen was tested as a mimetic of the key antioxidant enzyme glutathione peroxidase (GPx), a plethora of organoselenium compounds have been synthesized and tested for various pharmacological purposes. Here a brief overview of the most important achievements in bioactive organoselenium small molecules is given, with particular emphasis on the GPx-like as well as to the antiviral, antibacterial, antifungal, and antiprotozoal activities. While historical information is given to help contextualize the content, the most recent literature is comprehensively discussed.

2.1 General Introduction: Historical Aspects and Focus on Ebselen

The history of the biological properties of selenium (Se) mirrors the uncommon nature of this element. Indeed, Se is the sole element for which the toxicity, which can be considered as an exacerbation of the pharmacological activity, was reported before the element was actually disclosed. The Swedish chemist Jöns Jacob Berzelius officially discovered Se in 1817 at Stockholm, even if modern researchers guess that Arnaldo from Villanova originally observed Se many years earlier, in the 1300s. The toxicity of Se seems to be reported long before its discovery, precisely in 1325, by Marco Polo. The Italian trader reported, in his traveller journal, of a deadly poisoning affecting animals during his passage through heathlands of the Shaanxi province in China [1]. He reported that horses and beast of burden, after feeding on plants growing in that area, started losing their hooves. Modern researchers believe that the livestock disease described by Polo was selenosis due to the high levels of selenium in plants grown in those seleniferous soils. Interestingly, Marco Polo's selenosis was recently questioned by Chinese researchers who proved that the animal poisoning was caused by the eating of plants belonging to *Oxytropis* species. These vegetables contain some alkaloids able to provoke a clinical state similar to that of selenosis [2]. However, whether it has been selenosis or not, these circumstances created a peculiar scenario for elements

© Springer International Publishing AG, part of Springer Nature 2018
E. J. Lenardão et al., *New Frontiers in Organoselenium Compounds*,
https://doi.org/10.1007/978-3-319-92405-2_2

in the periodic table, the scenario in which the biological properties of the element were documented before its discovery.

In some cultures, selenium is linked to madness; the legend springs from locoweed, a variety of plants of West America. Such plants need lots of selenium to grow and bloom beautifully in the spring, attracting the attention of several animals like sheep and horses. When animals eat locoweed they develop a disease, which is, as the name of the plant suggests, similar to madness ("loco" means crazy in Spanish). Some storytellers attributed the General Custer's disastrous defeat at the Battle of the Little Bighorn in 1876 to the poisoning effect these plants exerted on the troops' horses.

A further folk story about selenium toxicity was the British beer poisoning in 1900 that caused 1000 deaths and was attributed to selenium and arsenic toxicity, both elements found at high levels in brewing sugar [3]. Of course, all these facts can be connected with the ability of some plants to extract selenium from so-called seleniferous soils, which was proven in 1930, almost simultaneously to the discovery of the toxicity of Selocide, $(KNH_4S)_5Se$ [4], the first systemic insecticide that have been approved [5]. Given the wide distribution of Selocide, the USA-Food and Drug Administration (FDA) commissioned some studies where it was found that toxic levels of Se, as Selocide or seleniferous wheat, induced liver cirrhosis, followed, after 18 months, by some benign tumors [6]. This led, by misinterpretation, to the concept that Se caused cancer, fuelling the fear for not only the element, but also for all of the Se-containing compounds. Because of this, the term "Selenophobia" was conceived by Douglas V. Frost in his seminal paper of 1972 [7]. After having detailed the key facts, the author stated that "*The ability to be right about Se may prove to be far more important than the ability to place a man on the moon*", highlighting how intricate was the information about the biological effects of selenium, at that time. After "Selenophobia," in 1988 "Selenophilia" was reported as the strong interest in the use of selenium for the prevention, alleviation or cure of a variety of disorders, which have not been shown to be directly associated with Se supplementation [8]. This made the relationship between selenium and cancer even more complicate. In fact, at various time points in the recent history, Se has been hypothesized to be either a cancer-inducing agent or a preventive element or a chemotherapeutic and, at the same time, to have no effect on human cancer [9].

Randomized controlled trials were carried out to determine the real relationship between selenium and cancer. Among them, the Nutritional Prevention of Cancer (NPC) Trial, the Selenium and Vitamin E Cancer Prevention Trial (SELECT), and more recently, SELEnium and BLAdder Cancer Trial (SELEBLAT) are worth mentioning. The NPC trial was conceived to prove the hypothesis that dietary Se supplementation may decrease cancer risk and it was conducted by administering 200 μg/day to about 1350 participants [10]. This study concluded that, while the administration increased the risk of total nonmelanoma skin cancer, the risk of all cancers including prostate, lung and colorectal cancers was decreased. The SELECT trial attempted to evaluate the activity of Se combined with vitamin E for prostate cancer prevention by recruiting more than 35,000 men [11]. Overall, the results of the SELECT trial provided evidence of no cancer-prevention effect in the studied population. After the conflicting results obtained with these two studies, SELEBLAT was meant to determine whether Se could reduce the recurrence of bladder cancer on 277 patients, and no correlation was finally found [12].

The suggestion of opposite effects of Se on health is not confined to cancer; indeed Se has been shown to be both neuroprotective and neurotoxic [13], a cardiovascular health promoter, as well as a cardiovascular health risk factor [14], and has been suggested to be both an antidiabetogenic and a prodiabetogenic agent [15]. Recently, new data have been generated and discussed by Vinceti, who highlighted how debated such relationship is [16].

The good face of Se was not fully shown until the late 1950s, when its role in the nutrition of many species was discovered by the seminal work of Schwarz and Foltz, who first established its importance as an *essential trace element* involved *in oxidation–reduction reactions* [17], and Pinsent, who reported that Se is required for optimal enzymatic properties of intestinal *Escherichia coli* [18]. A series of observations highlighted how selenium deficiency is linked to several diseases, such as white muscle disease in cattle and sheep [19], exudative diathesis in chicken [20], and male infertility in mammals [21].

Another milestone in the history of Se was the discovery that the antioxidant enzyme glutathione peroxidase (GPx) is actually a selenoenzyme [22–25]. In the 1970s, more selenoproteins, such as the protein A component of the clostridial glycine reductase [26], formate dehydrogenase [27], nicotinic acid hydroxylase, and xanthine dehydrogenase [28] were identified. Se is present in all of these proteins in the form of the amino acid selenocysteine, which was identified in 1975 [29–31]. Up to now, 25 selenoproteins are known, some of them play roles that are still unknown [32].

The 1980s were important since selenium, for the first time, was considered beneficial for human health [33, 34]. These years were momentous since medicinal chemistry efforts appeared in literature. In 1983, selenazofurin was reported as an antiviral [35], and in 1984, the ability of an organoselenium compound, named ebselen **1** (Fig. 2.1), to act as GPx-mimic was reported [36]. The seminal work was carried out by Helmuth Sies. Starting from this pillar, a whole branch of research was, and continues to be, devoted to the identification of compounds able to reproduce the GPx catalytic activity [37, 38]. Within this chapter, a special section will be devoted to the organoselenium compounds endowed with GPx-mimic activity. The research work carried out in 1984 by Sies was also important, because it paved the way for the investigation of the pharmacological properties of ebselen. Ebselen was synthesized by Lesser and Weiss in 1924 [39] and was considered irrelevant from a pharmacological point of view for exactly 60 years. From 1984, nearly 1000 articles focused on its biological properties have been exponentially reported, as depicted in Fig. 2.2, which was obtained by interrogating in the PubMed engine typing "ebselen" as the subject.

Starting from 1998, ebselen was submitted to several clinical trials (summarized in Table 2.1) and after 20 years of intensive human testing, it was included into the NIH clinical collection (NCC), which is a library made of 727 compounds with a history in phase I–III of clinical trials. This collection gives the possibility to screen

Fig. 2.1 Noteworthy facts in the history of selenium

compounds endowed with the suitable toxicity and pharmacokinetic properties to straight become drugs. By screening the NCC, ebselen emerged twice as hit compound in two recent papers, in particular it was found to attenuate depression and mania in a lithium-like fashion [40] and to exert anti HCV activity by inhibiting NS3 protease (we will back to this topic later in this chapter) [41]. Both activities were ascribed to the compound ability to chemically react with key cysteines.

The readers who are interested in the development of ebselen are directed to Ref. [48], in which it is possible to find information about its discovery and development both from an academia and from an industrial point of view. This article is authored by Sies, which gave his personal perspective of the way the research on this fascinating compound unfolded.

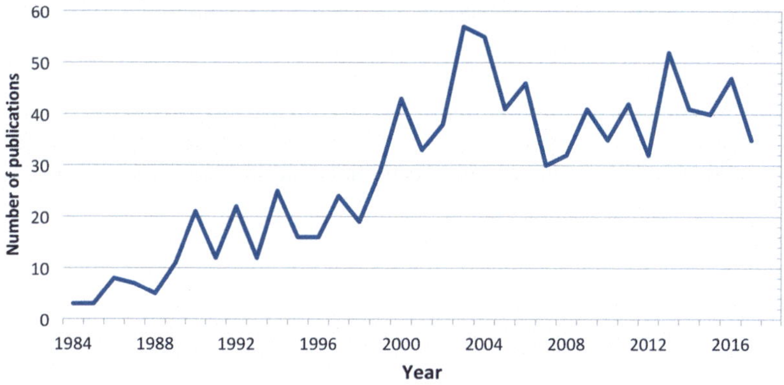

Fig. 2.2 Number of publications focused on ebselen **1** starting from its discovery

Table 2.1 Clinical trials carried out with ebselen **1**

Year	Title/objective	Patients	Ref
1998	Ebselen in acute ischemic stroke: a placebo-controlled, double-blind clinical trial. Ebselen study group	300	[42]
1998	Neuroprotective effect of an antioxidant, ebselen, in patients with delayed neurological deficits after aneurysmal subarachnoid hemorrhage	286	[43]
1999	Ebselen in acute middle cerebral artery occlusion: a placebo-controlled, double-blind clinical trial	99	[44]
1999[a]	Signalling mechanisms and vascular function in patients with diabetes mellitus	60	NCT00762671[b]
2009	Development of ebselen, a glutathione peroxidase mimic, for the prevention and treatment of noise-induced hearing loss	32	[45]
2011[a]	Otoprotection with SPI-1005 for prevention of temporary auditory threshold shift	83	NCT01444846[b]
2015[a]	Study to evaluate SPI-1005 in adults with Meniere's disease	40	NCT02603081[b]
2016	Effects of the potential lithium-mimetic, ebselen, on brain neurochemistry: a magnetic resonance spectroscopy study at 7 tesla	20	[46]
2016	Ebselen does not improve oxidative stress and vascular function in patients with diabetes: a randomized, crossover trial	26	[47]
2017[a]	Ebselen as an add-on treatment in hypo/mania	60	NCT03013400[b]
2017[a]	SPI-1005 for prevention and treatment of aminoglycoside induced ototoxicity	100	NCT02819856[b]
2018[a]	SPI-1005 for prevention and treatment of chemotherapy induced hearing loss	80	NCT01451853[b]
2018[a]	A phase IIb study of SPI-1005 to reduce the incidence, severity, and duration of acute noise induced hearing loss (NIHL)	180	NCT02779192[b]

[a]The results of these studies have not been published
[b]The given reference is the code of the clinicaltrials.gov website, which identifies the cited study

Ethaselen **2** (Fig. 2.1), is the second organoselenium compound to be submitted to clinical trials for the treatment of non-small cell lung cancers overexpressing thioredoxin reductase (TrxR) [49]. Currently, it is in clinical trial phase II [50]. Ethaselen was designed to inhibit mammalian TrxR activity by selectively targeting SeCys498/Cys497, the C-terminal active sites of mammalian TrxR, with high binding potential [51, 52]. Furthermore, TrxR inactivation correlates with cell death/apoptosis in a plethora of cell lines [50].

The last organoselenium compound under clinical evaluations is the selenide ALT2074 **3** (Fig. 2.1), also known as BXT-51072. It was identified as a GPx-mimic able to prevent endothelial alterations and myocardial ischemia–reperfusion injury [53, 54] and currently it is the subject of two phase II clinical trials. The first one is meant to assess the safety, the pharmacokinetic profile and characterize the effect on biomarkers of inflammation and oxidative stress in diabetics, with evidence of coronary artery disease (study number NCT00491543) as a result of the promising in vivo preclinical evaluation [55]. The second one is meant to determine the cardioprotective effect of selenide **3** in diabetic patients undergoing elective angioplasty (study NCT00320502).

Due to the recent publication of excellent reviews, books and book chapters [56–58], the present chapter will not cover all the literature of the biological activities of organoselenium compounds; it will be instead focused on the most recent advances on GPx-mimics and antimicrobial agents.

2.2 Selenium-Based GPx-Mimics and Antioxidants

The development of selenium-based antioxidants is tightly connected with that of the compounds able to mimic the catalytic cycle of the key antioxidant enzyme GPx. Although the concepts of "antioxidant" and "GPx-mimic" have been often used as synonyms, they should be viewed and studied from different perspectives. In the authors' opinion, a "GPx-mimic" is a compound able to catalyze the oxidation of thiols to the corresponding disulfides in the presence of an oxidant, and nothing more. From this standpoint, it comes clear that the measurement of the GPx-mimic activity of a given compound corresponds to the measurement of a chemical instead of a pharmacological property. As a result, a good GPx-mimic activity measured in vitro could not be translated in an antioxidant capacity in vivo or in a cellular context. Rather, it is common that a good GPx-mimic does not behave as an antioxidant in living systems, but even as a pro-oxidant. In addition, besides the GPx-mimic properties, the antioxidant activity of a selenium-containing compound can be the result of its ability to scavenge reactive oxygen species (ROS) or of its metal binding characteristics, as proved by several reports that will not be detailed in deep (for a detailed discussion the interested reader is directed to Ref. [59]).

In this chapter, the GPx activity will be discussed along with the methods developed for its evaluation, and, where possible, the structure–activity relationship

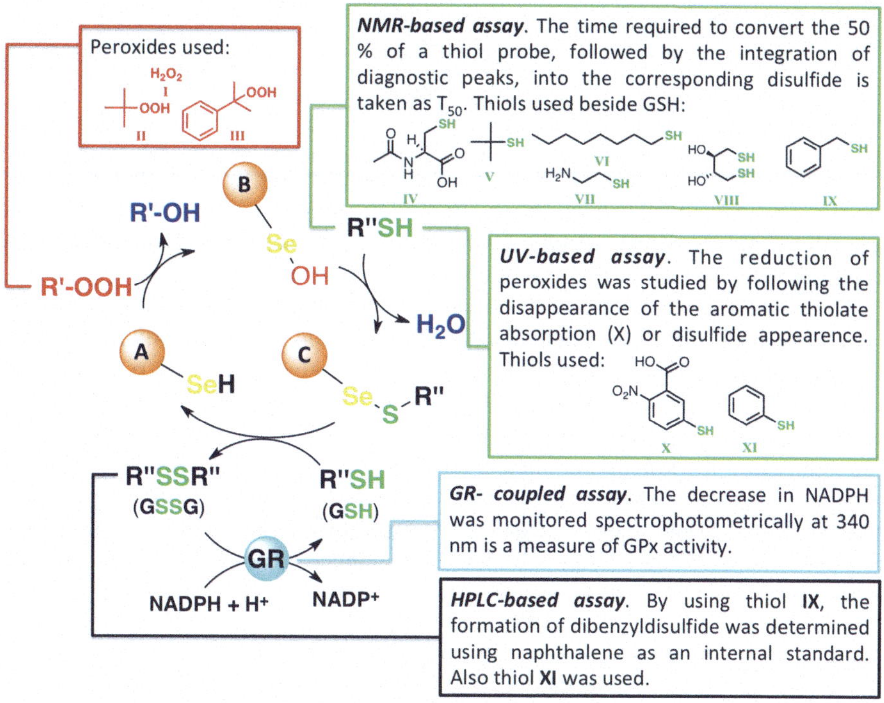

Fig. 2.3 GPx mimic cycle and the assays developed for its evaluation

(SAR) will be detailed in order to give fruitful insight for the next generation of GPx-mimic compounds.

The history of GPx mimics can be traced back to the 1970s, when it was discovered that the key antioxidant enzyme GPx was actually a selenoenzyme. In 1973, Rotruck [22] hypothesized the presence of selenium as an integral part of the enzyme, while Flohè [23] proved that the chalcogen was present in the protein in stoichiometric amounts. As a result, Flohè is recognized as the father of such an important discovery. Successively, it was discovered that selenium takes part in the structure of the enzyme as selenocysteine (SeCys) [60]. Nowadays, eight isoforms of GPx are known, with GPx1 being the most abundant [61]. GPx1 removes from the cellular environment dangerous peroxides, such as hydrogen peroxide (H_2O_2) or organic peroxides (ROOH), following the catalytic cycle showed in Fig. 2.3. In particular, selenocysteine (**A**), which in the active site forms a "catalytic triad" with tryptophan and glutamine, reacts with peroxides, and is transformed into selenenic acid (**B**), which is attached by a molecule of reduced glutathione (GSH) freeing water and leading to the selenylsulfide intermediate (**C**). Another molecule of GSH gives back the intact enzyme and GSSG, which is reduced back to GSH by glutathione reductase at the expenses of NADPH (not shown) [62].

Considering that in the 1970s the ability of some selenium compounds to decompose hydroperoxides and oxidize thiols to disulfide was already known [63, 64], it

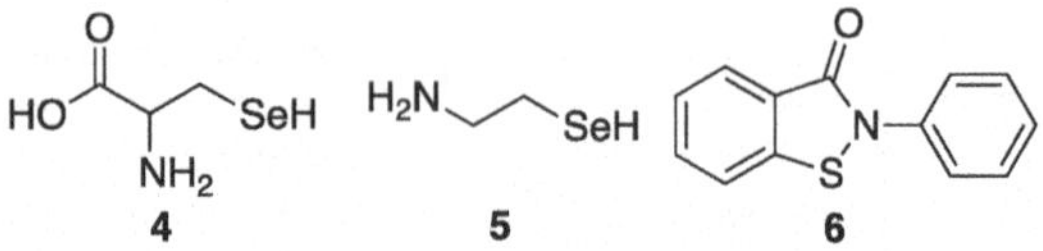

Fig. 2.4 Structures of compound **4–6**

comes natural for scientists to assay whether this property would follow the same mechanism as that of the newly discovered selenoenzyme (GPx). During the years several methods were developed to assay the ability of selenium compounds to act as pseudo enzymes, all of them are summarized in Fig. 2.3.

The glutathione reductase (GR) coupled assay was the first method used for the evaluation of the GPx-mimic activity [65, 66]. It is considered an indirect method. The glutathione (GSSG) formed as a result of the catalytic activity is reduced back to GSH thanks to the activity of the GR enzyme at the expenses of NADPH. The decrease of NADPH, followed spectrophotometrically at 340 nm, is taken as read-out to measure the kinetic parameters and thus the GPx mimic activity. Different peroxides were employed; for example, the activity of D,L-selenocystine (compound **4**, Fig. 2.4) and selenocysteamine (**5**) was evaluated in the presence of hydrogen peroxide, *tert*-butyl hydroperoxide, and cumene hydroperoxide. The results were expressed in terms of kinetic parameters, such as the Michaelis–Menten constant (Km), which gives an idea of the affinity for the substrate, and maximum reaction rate (V_{max}). In all the cases the results were similar; when compared with the native enzyme, it was found that compounds **4** and **5** were endowed with a lower affinity for both GSH and the peroxides [66].

Sies et al. measured the GPx-mimic activity of ebselen (**1**) and of its sulfur analogue (compound **6**) by following the disappearance of NADPH at 366 nm [36]. The same method was later applied by Wilson to test, for the first time ever, the GPx-like activity of some diselenides [67]. A prerequisite condition for obtaining reliable results with this assay is that all intermediates in the reaction are inert toward GR and its cofactor (NADPH). This is not an easy task especially for those compounds whose mechanism of the GPx-like behavior is unknown. As example, it was proven that the selenylsulfide formed from the ebselen and GSH reaction serves as a substrate for GR, thus hampering the activity to be properly evaluated [68]. For this reason, direct methods, able to directly monitor the consumption of one of the reagents, were developed. In this context, 3-carboxy-4-nitrobenzenethiol (compound **X**, Fig. 2.3) was used as thiol probe and the kinetic of the reaction was obtained by following its consumption as a decrease of the absorbance at 410 nm [69]. Later on, Iwaoka and Tomoda developed a UV-based assay in which the reaction parameters were extrapolated by using benzenethiol (compound **XI**) as a probe, monitoring the increment of absorbance at 305 nm as a result of the formation of diphenyl disulfide [70].

As a direct method, the degradation of H_2O_2 can be determined by following its concentration versus time, according to the Hildebrandt and Roots method, based on the H_2O_2-mediated oxidation of Fe^{2+} to Fe^{3+} in turn complexed to thiocyanate and quantified spectrophotometrically at 480 nm [71]. This method was used for the evaluation of ebselen using as thiol probe, besides GSH, lipoic acid [72]. NMR-based assays were developed starting from the seminal work of Engman, who moni-

Table 2.2 Thiol probes used in NMR-based assays and their T_{50} in the absence of catalysts

Probe	T_{50} (h)	Solvent	Ref
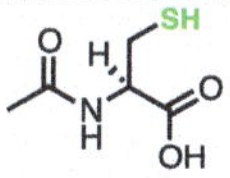 **IV** *N*-Acetylcysteine	55	D_2O/CD_3OD (4/1)	[68]
 **V** *tert*-Butyl mercaptan	≫100	CD_3OD	[68]
 **VI** 1-Octyl mercaptan	≫100	CD_3OD	[68]
 **VII** Cysteamine	≈2	CD_3OD	[73]
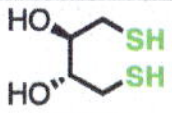 **VIII** Dithiothreitol (**DTT**)	≫5	CD_3OD	[73, 74]
	>400 s	D_2O	[75]
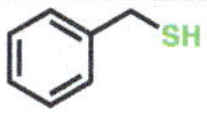 **IX** Benzyl mercaptan	≈3	CD_3OD	[73]
GSH		D_2O DMSO	[76]

tored the conversion of thiol to disulfide by the integration of key peaks in the 1D ^{1}H NMR; the integral area was then plotted against time to give a linear correlation. The time required to convert 50% of the thiol probe has been referred to as T_{50}, which is a valuable tool, nowadays widely used, to rank and compare different GPx-mimic compounds. The most important aspect in this approach is the choice of the thiol probe, which needs to be selected among those whose spontaneous oxidation to disulfide is slow in the presence of oxidants but in absence of the catalyst. As example, GSH is not exploitable since its oxidation in water containing H_2O_2 at pH = 7 is complete in 15 min [72]. The thiol probes that have been used in the NMR-based assays are summarized in Table 2.2.

N-Acetylcysteine (**IV**), *tert*-butyl mercaptan (**V**), and 1-octyl mercaptan (**VI**) react very slow in the absence of catalysts, while cysteamine (**VII**), DTT (**VIII**), and benzyl mercaptan (**IX**) showed faster kinetics. DTT was also used in deuterium oxide rather CD_3OD, affording very fast kinetics, almost completed in minutes time lapses.

In 1997, an HPLC-based method was proposed; benzyl mercaptan (**IX**) was used as thiol probe and the time-dependent formation of dibenzyl disulfide was followed by using naphthalene as an internal standard [77]. Moreover, benzenethiol (**XI**) was used in an HPLC-based method able to measure the formation of diphenyl disulfide, which was also used as internal standard [78].

Each method has some variants leading to a very complex scenario made of a huge amount of results quite difficult to interpret in an objective manner. In the following sections, the GPx-activity of the organoselenium compounds will be detailed according to their chemical structure.

2.2.1 Ebselen and Related Structures

Ebselen **1** was tested in almost every model of GPx-mimic assays and little differences in terms of operative details in the same method led to substantial differences in the final results making the comparison of the different methods as well as different compounds a very hard task. As stated, the first experiments were carried out by Helmuth Sies in 1984, who tested the compound in the GR-coupled assay highlighting how the kinetic of the catalytic reaction is saturated upon the addition of 2 mM of GSH, while the speed of GPx catalyzed reaction increased linearly with the concentration of GSH [36]. In addition, Sies tested the antioxidant capacity of compound **1** in the microsomal lipid peroxidation model. In the same year, Wendel measured the kinetic parameters of the ebselen-catalyzed reaction by different methods in the presence of various peroxides and compared them with the native enzyme [79]. In addition, they found that the methylation of selenium drastically reduced the GPx-mimic activity. In these years scientists proposed that ebselen was a sort of selenium reservoir for GPx, because it was found to restore the enzymatic activity in rat fed with a selenium-deficient diet but supplemented with the compound. Besides the GPx-mimic activity, ebselen behaves as an antioxidant because it protected rat liver microsomes [36] and isolated hepatocytes [80] against ascorbate/Fe-induced lipid peroxidation. Besides GSH, other thiols were also employed to assist the ebselen-mediate reduction of hydrogen peroxide, such as *N*-acetyl cysteine (**IV**) [81], dithiothreitol (**VIII**) [80], and dihydrolipoate [72].

Ebselen was assayed in a series of HPLC-based experiments where it was introduced the concept of the half-life of the oxidations ($T_{1/2}$, also referred to as T_{50}), which is the time required for the 50% conversion of a thiol probe into its corresponding disulfide. In the HPLC experiments, the thiol probe was benzenethiol **IX** (Fig. 2.3 and Table 2.2), that was catalytically converted by ebselen into its disulfide, using ^{t}BuOOH as the oxidant, in a T_{50} of 42 h. Actually, in these experiments ebselen was used as a reference compound, and it was found to be the least active among the tested compounds [82–84]. Mugesh used compound **XI** and its tolyl analogue as thiol cofactors in a series of HPLC-based GPx mimic analysis, experiencing, for the first time, the thiol exchange reaction, which blocks the catalytic cycle and thus reduces the catalytic capability of the organoselenium compound [85].

It is known that heteroatom-containing groups in the active site of GPx or similar groups in model compounds interact with selenium modulating its reactivity. In some cases, these interactions have been shown to increase the electrophilic character of selenium and when established in the selenylsulfide intermediate (**C** in

Fig. 2.5 Thiol exchange reaction in the selenylsulfide intermediate

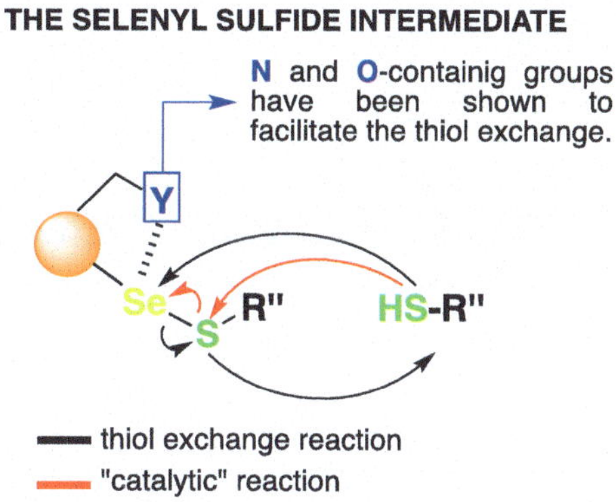

Fig. 2.3) of the pseudo enzymatic cycle, they become detrimental. In particular, strong Se---*N* and Se---*O* interactions enhance the nucleophilic attack of the incoming thiol at the selenium atom instead of the desired attack at sulfur, leading to a thiol exchange reaction, also called *ping-pong* reaction, that hampers the regeneration of the active species selenol and thus blocking the catalytic cycle (Fig. 2.5) [85].

From a general point of view, ebselen is a poor catalyst when the thiol partner is an aromatic compound, while it exerts a significant catalytic activity when the substrate is GSH [86]. This could explain why ebselen is a poor GPx-mimic compound while it exerts a pronounced antioxidant activity in vivo which, however, strongly depends on the metabolic status of the cell the antioxidant activity is tested in. Indeed, it was shown that ebselen can act either beneficially or detrimentally through the depletion of GSH [87]. In the presence of a high concentration of GSH, the effect of ebselen will be primarily beneficial, while in a model in which GSH concentration is low, the depletion of GSH may dominate over the antioxidant activity. Besides the GPx-mimic activity, other pharmacological properties may account for its antioxidant activities observed in living systems; for example, the ability to increase the pool of thioredoxin, by inhibiting TrxR [88], or the rapid reduction of peroxynitrite [89], was found to be of pivotal importance for the antioxidant properties of the compound in a cellular environment. Noteworthy, ebselen is completely devoid of any toxic effect, making it the most safe organoselenium compound known to date, as confirmed by several studies and clinical trials [90]. This is probably because elemental selenium is not easily released from the ebselen structure.

As a positive control, ebselen was tested in UV-based experiments using compound **XI** as thiol probe and hydrogen peroxide as oxidant and it was able to catalyze the disulfide formation with a T_{50} of 155 min, which was sensibly lower than that showed by diselenides [91]. In NMR-based experiments, a T_{50} of about 60 min was reported [92, 93].

Alongside the antioxidant, other pharmacological properties are worth to be mentioned for compound **1**: it inhibits divalent metal transporter 1 and the phosphorylation of the tau protein involved in Alzheimer's disease (AD). This particular behavior was exploited in the frame of medicinal chemistry projects in which the benzisoselenazolone pharmacophore was merged with known marketed anti-AD drugs for the development of designed multiple ligands [94]. In a recent report, it

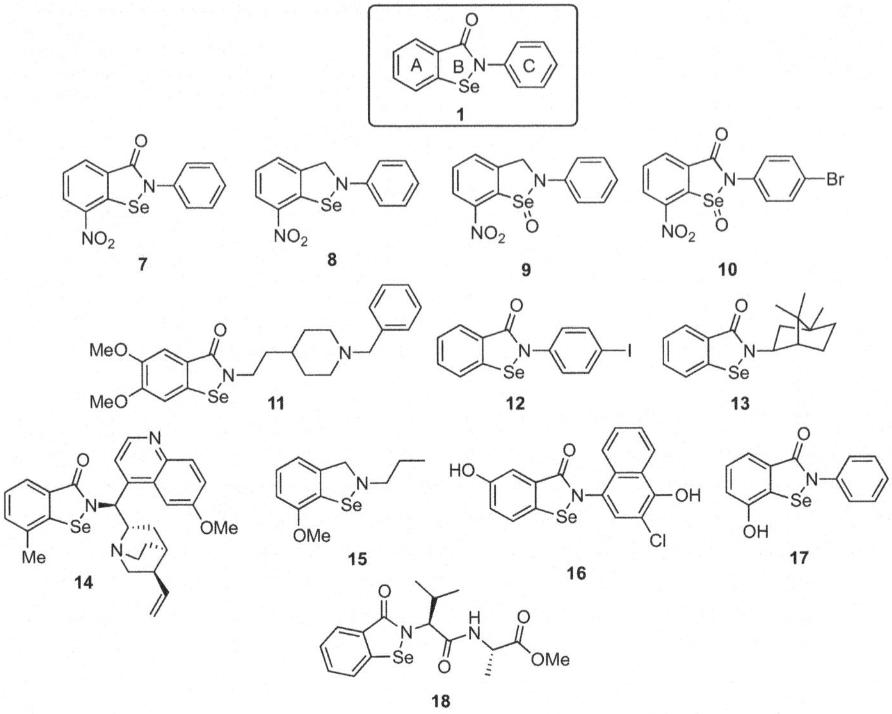

Fig. 2.6 Ebselen and related compounds

was able to block inositol monophosphatase (IMPase) and thereby induce lithium-like mood-stabilizing effects in mice [40]. Ebselen was found to be a skin depigmenting agent by inhibiting the biosynthesis of melanin and melanosomal transfer [95]. Recently, ebselen has been considered as a novel therapeutic candidate for the treatment of type 2 diabetes because of its ability to ameliorate fasting hyperglycemia and reduce the deterioration of β-cell mass and function [96], and to improve glucose-stimulated insulin secretion in murine islets [97].

Using compound **1** as a template, a plethora of compounds were designed, synthesized and tested for their ability to mimic GPx [38, 98]. They will not be comprehensively collected here, but a series of analogues chosen among the most recent examples will be reported by emphasizing the structural determinants able to impart a strong GPx mimic activity, and, when tested, the antioxidant activity.

One of the most fruitful modifications aimed to improve the GPx-mimic activity of ebselen was the introduction of a nitro group in the *ortho* position at the selenium atom (compound **7**, Fig. 2.6), most probably because the presence of such a group weakens the Se---*O* interaction in the selenylsulfide intermediate, preventing the thiol exchange reaction. The presence of the *ortho* nitro group was maintained in the successive series of GPX-mimics, characterized by the presence of methylene in the place of the carbonyl group (compound **8** and its selenoxide analogue **9**) and a selenoxide coupled with a phenyl substituted with bromine at the *para* position (compound **10**). All of the ebselen analogues **8–10** showed an improved catalytic efficiency with respect to compound **1** measured in terms of initial rate (V_0) with the GR coupled assay [99]. Derivative **11**, named "selenpezil," was developed by hybridizing the pharmacophore of ebselen with that of donepezil, an acetylcholinesterase (AChE) inhibitor used for the treatment of AD. Compound **11** exerted a GPx-mimic activity superior to that of the parent compounds and was also endowed with the ability to inhibit AChE and prevent plaque aggregation, which is a key point in the development of drugs acting against AD [100].

A series of analogues were prepared by modifying the ring C of ebselen, and according to NMR measurements, the most potent compounds are the iodine analogue **12** [101] and the bornyl derivative **13** [92]. The quinine derivative **14** showed the unique ability to generate the corresponding selenol upon reaction with compound **XI**, which is of pivotal importance for the proper beginning of the catalytic cycle. This particular behavior, which differentiates compound **14** from ebselen and other tested compounds, reflected a very good GPx-mimic activity measured by following the PhSSPh formation through UV spectroscopy [102]. The *N*-propyl substituted isoselenazole **15** showed an improved thiol-oxidizing activity coupled with antioxidant activity in different cell lines [88]. A SAR insight can be obtained from the GPx evaluation of compound **15** and **8**; for the isoselenazole class of compounds, the electronic properties of the group connected to the selenium at the *ortho* position is not important, since both electron withdrawing and donating groups gave compounds with pronounced GPx-mimic properties. A computer-aided design of GPx-mimic compounds was performed leading to the synthesis of compound **16**, which is characterized by the presence of a hydroxyl group at the selenium *para* position. It exerted an outstanding activity because of a very weak non-covalent Se--*O* interaction in the selenylsulfide intermediate [103]. Finally, the introduction of the hydroxyl group led to compound **17**, which showed a catalytic efficiency 15-fold higher than ebselen [104].

A series of dipeptide-containing ebselen analogues were prepared and tested for their GPx-like activity through the GR method using compounds **I–III** as peroxide sources. An important feature to highlight is that the activity of the synthesized compounds is highly dependent on the peptide attached to the nitrogen atom and the peroxide used. Compound **18**, bearing a Val-Ala peptide, showed the best activity in the series among the three peptides used. The reason behind the high activity of compound **18** is the facile generation of the corresponding selenol upon reaction with two molecules of GSH, thus indicating that the thiol exchange reaction is practically absent for this compound (Fig. 2.5). Compound **18** strongly inhibited the peroxynitrite-mediated nitration of tyrosine [105]. Unfortunately, this compound was never tested in a cellular model to assess whether the GPx-mimic and anti-peroxynitrite activity measured in vitro translates in an antioxidant ability in a living contest.

The chemical mechanism through which all of these compounds exert the GPx catalytic activity is still matter of debate and it may vary according to the substitutions on the isoselenazolone scaffold. In this chapter, such mechanisms are not going to be discussed since it was the precise object of a book chapter recently published by some of us [37].

As ebselen, its analogues are endowed with pharmacological properties that go beyond the antioxidant one. As example, ebselen oxide has been described as a selective inhibitor of α-methyl-acyl-CoA racemase, which converts dietary (*R*)-branched fatty acids to the natural (*S*)-enantiomers [106]. Other ebselen analogues were reported as anticancer [57], antibacterial and antiviral agents (see later in this chapter) among other activities.

2.2.2 Diselenides

Together with ebselen and its analogues, diselenides are one of the most studied class of GPx-mimic compounds. Even if in many articles it is reported that diselenides are more potent than isoselenazolones as GPx-mimics, in the authors' opinion it is impossible to draw a general conclusion. The extreme variability of the conditions under which the GPx-mimic activity has been determined makes a statistically significant and general comparison impossible.

The simplest compound belonging to this class is diphenyl diselenide (compound **19**, Fig. 2.7), which was tested in several models as a reference and resulted a better catalyst than ebselen, probably because of the lack of intramolecular interactions in the context of the selenylsulfide intermediate. The mechanism by which diselenides exert the GPx-like activity is similar to that of ebselen and analogues, entering the catalytic cycle after the nucleophilic attack by the thiol probe, which generates benzeneselenol in the active site of the enzyme. For compound **19**, this mechanism was demonstrated in 1994 and it was important because, since that time, the Se---*N* bond was thought to be essential for the GPx-mimic activity [107, 108].

Several substitutions have been pursued taking as a template the diphenyl diselenide scaffold. The phenyl ring of **19** was replaced by the pyridine one in compound **20**, which was more potent than **19** in two separate assays. In addition, the pyridine analogue presented antioxidant activity by the complexation with radical-generating metals [109]. 2,2′-Dipyridyl diselenide **20** had a higher in vitro antioxidant effect and lower inhibitory effect on δ-aminolevulinic dehydratase (δ-ALA-D) activity than other disubstituted diaryl diselenides in rat liver homogenates [110]. δ-ALA-D is a key human enzyme, which has been reported as sensitive to pro-oxidants and heavy metals and its inhibition is considered a marker of toxicity [111]. Then, the pyridine pharmacophore was merged with the strong antioxidant pyridoxine (not shown), because of its high reactivity towards singlet oxygen, leading to compound **21**. By using the GR coupled assay, compound **21** was found to be endowed with a superior catalytic activity compared to ebselen [112]. The catalytic ability of **21** was even increased by the addition of a bromine atom in the pyridine ring [113]. Bis-nicotinoylamide diselenides (as an example, compound **22**) were prepared in an attempt to reproduce the biological properties of vitamin B3 and nicotinamide, that are important biomolecules involved in a wide variety of biological processes,

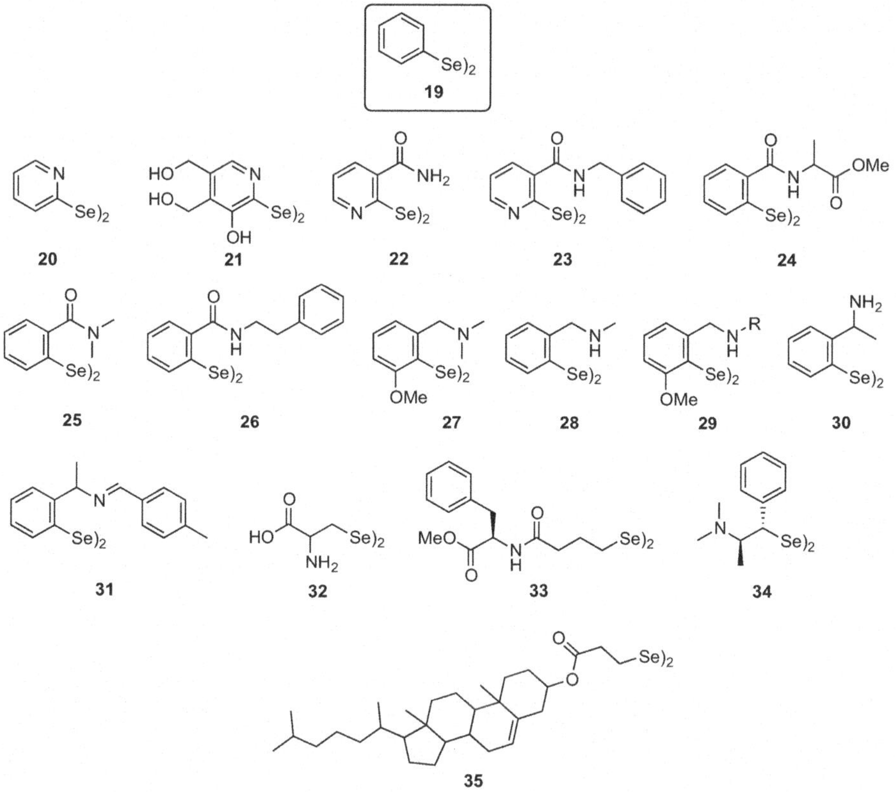

Fig. 2.7 Structures of diorganyl diselenides

including energy supplying, synthesis of fatty acids, besides showing antioxidant properties. Compound **22** exhibited a potent GPx-mimic activity measured by NMR and HPLC-based assays, these activities being well correlated with the antioxidant activity measured in cellular models [114, 115]. The mechanism by which pyridine-containing compound **22** or **23** mimic the catalytic cycle of GPx is depicted in Scheme 2.1. Firstly, diselenide reacts with GSH to form the selenylsulfide intermediate (**E**) and selenol (**D**). The more stable tautomer of **D** is the selenone **F**, which can be also formed by the reaction of **E** with a further molecule of GSH. Then, the peroxide triggers the formation of selenenic (**G**) and seleninic (**F**) acids, which are capable to oxidize further molecules of GSH.

Diselenide **23** was designed and prepared in the frame of a wider study in which the picolinamide fragment was inserted into either aromatic or aliphatic derivatives. The aromatic compound showed a high activity, even higher than diphenyl diselenide **19**, which was used as a reference compound. The GPx mimic activity was coupled with a pronounced inhibition of the production of thiobarbituric acid reactive substances (TBARS) measured in brain homogenates [116]. Several amide-based diselenides were prepared by introducing amino acids and, in particular the

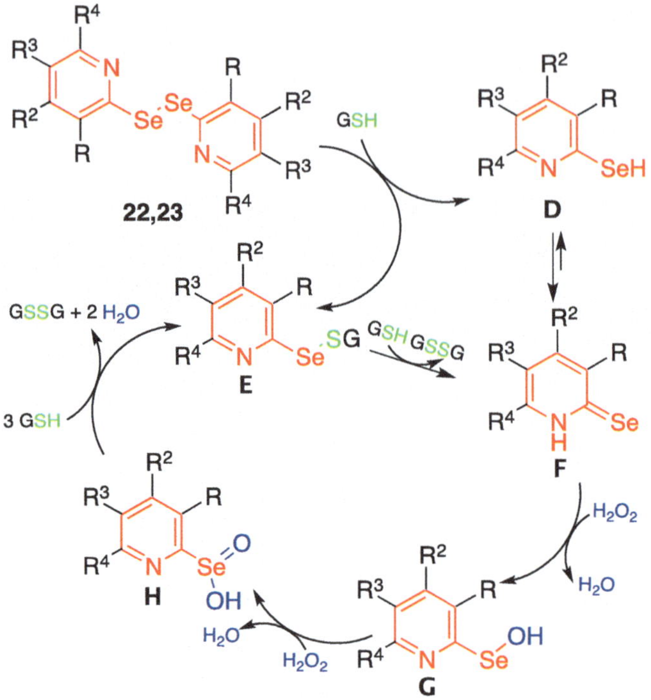

Scheme 2.1 Catalytic cycle of pyridine-containing diselenides

alanine derivative **24**, showed a V_{max} that was noticeably higher than that of ebselen and its analogues [117]. Earlier, it was reported that tertiary amides, such as **25**, were much better catalysts than the corresponding secondary ones in the reduction of peroxides by thiols, and this was explained by the weakening of the Se---*O* nonbonding interactions. Another interesting finding was that the activity of the tertiary amides was independent of the peroxide used, differently that the observed for dipeptide-containing ebselen-like structures and secondary amides, whose activity varied using different peroxides [118]. Amphetamine-containing amide-based diselenides were prepared, as example compound **26**, in order to check if the GPx-mimic activity—measured through UV measurements using PhSH **XI** as a probe, would correlate with the observed protection against lipid peroxidation and acting thus as substrates for TrxR [91].

Ortho amine-containing diselenides were rationally designed and synthesized aiming to improve the GPx-like properties. Indeed, the presence of the basic terminal unit would promote the activation of the selenol intermediate (RSeH) into the selenolate anion (RSe$^-$), while the Se---*N* nonbonding interactions in the selenenic acid (RSeOH) would prevent the conversion of chalcogen into the more oxidized selenium species, while minimizing the nucleophilic attack of the thiol at the Se atom in the selenylsulfide intermediate, thus preventing the thiol exchange reaction [70]. Later, thanks to the experience gained with the ebselen-like structures, it was demonstrated that, to enhance the GPx-mimic activity, the Se---*N* nonbonded interaction should be weakened [119], for example by placing a methoxy group in the *ortho* position of the diselenide bridge, like in compound **27**. Intriguingly, the shift of the methoxy group to the *para* position reduced the catalytic activity, probably because of the lack of steric hindrance that, in the selenylsulfide intermediate, facilitates the incoming thiol to attack sulfur in the place of selenium [120]. Diselenides containing a secondary amino group, such as in compound **28**, were more active than the tertiary analogues [121]. Thus, by combining the structural features of compounds **27** and **28**, compound **29** was synthesized in 2015 and showed a noticeable catalytic activity [88]. Recently, the amino and imino derivatives, **30** [122] and **31** [123] respectively, were prepared and tested for their catalytic properties. Interestingly, both were more active than compound **19**, which was used as reference. While $(PhSe)_2$ **19** was able to oxidize thiols per se, without the addition of peroxides, derivatives **30** and **31** did not. The intrinsic oxidative property is deleterious and was linked to the major toxicity exerted by compound **19** in living systems [124].

The investigation of aliphatic diselenides started with the synthesis of selenocystine (compound **32**) and analogues, which were tested for their GPx-mimic activity, their affinity to peroxides and thiols and for their propensity to participate in redox reactions [125]. A series of aliphatic selenides and diselenides functionalized with amino acids were prepared and assayed by using compound **XI** as probe and following its oxidation to PhSSPh by UV spectroscopy. Among the tested compounds, the best in class was **33**, showing a T_{50} of about 40 min, lower than that showed by compound **19**, which was tested in parallel [126]. Similarly, the ephedrine-containing diselenide **34** showed a T_{50} value 11-fold superior than that exhibited by **19** [127]. A series of cyclopentaneperhydrophenantrene-containing diselenides (e.g., compound **35**) were prepared as liquid crystalline compounds to be used as functional organic materials in different research fields. The compounds showed an interesting GPX-mimic activity, even if they were tested in comparison with ebselen instead of diphenyl diselenide [128].

2.2.3 Selenides

Diorganyl selenides are another class of selenium-containing compounds widely studied for their ability to function as GPx-mimics, and a lot of compounds have been reported, starting in the 2000s [73, 74, 129, 130]. From a general point of view, the overall GPx-like activity of selenides is not as high as those of selenol-forming GPx mimics, such as diphenyl diselenide **19** and ebselen, basically because the reducing ability of selenides is lower than that of selenols; actually, exceptions to this general rule have been reported [131]. Here, just the most significant and recent examples will be discussed, starting from the work of Braga and co-workers, who clarified the mechanistic aspects through which this class of compounds exert the GPx-like activity. As depicted in Scheme 2.2, the selenide is initially oxidized

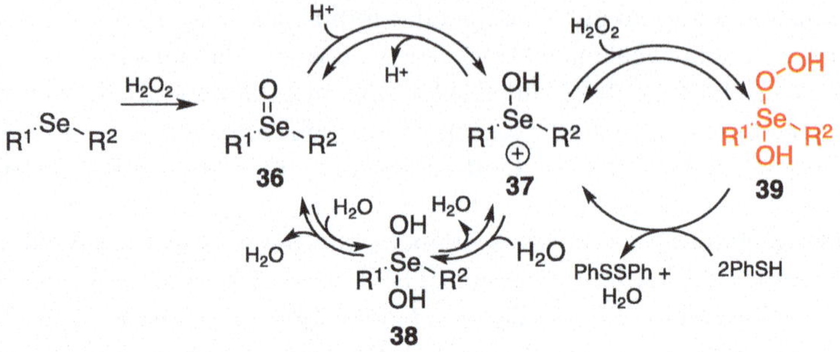

Scheme 2.2 Catalytic cycle of selenides proposed by Braga

to the corresponding selenoxide (compound **36**) that, in the presence of protons, can establish the equilibrium with the reactive selenonium **37**. Compound **37** can add a molecule of water, leading to **38**, or of hydrogen peroxide, leading to **39**, which is the actual catalyst able to oxidize benzenethiol to the corresponding diphenyl disulfide [132].

Iwaoka et al. investigated a series of *trans*-3,4-dihydroxyselenolanes and they observed the superior activity of the cyclic compound **41** (Fig. 2.8) with respect to its linear analogue **40**. The superior activity of **41** over **40** was confirmed in three experimental settings using mono-, di-, and poly-thiols as probes. The GR-coupled, the NMR and HPLC-based assays were employed, thus indicating that the observed catalytic activity is independent of the condition in which it has been tested. Such activity was ascribed to the cyclic structure, which elevates the selenium HOMO energy level, making it more exposed to the surroundings and more available for the reaction [74]. Compound **41** was later studied as a gastroprotective agent, being able to reduce the indomethacin-mediated stomach damage [133].

More lipophilic compounds were developed by using compound **41** as template in order to prepare antioxidants capable of reproduce the activity of GPx4 rather than that of GPx1. GPx4 is a phospholipid hydroperoxide glutathione peroxidase which differs from the other GPx isoforms because of its monomeric form (the others are tetrameric) and catalyzes the reduction of phospholipid hydroperoxides within membranes using water-soluble thiols as substrate [134]. One example is the myristyl derivative **42**, which proved to inhibit lipid peroxidation in a biphasic system. However, when measured with the GR-coupled assay, the GPx-activity of this compound is very low, probably due to its poor solubility in the buffer solution [135, 136].

Compounds **43–47** were prepared in the frame of a long-lasting project meant to elucidate the GPx-mimic mechanism of selenides and selenuranes [84, 137]. The compounds were evaluated with two experimental settings consisting of an HPLC-based and an NMR-based assays. Interestingly, in this latter assay for the first time GSH was used as the thiol probe in D_2O. Compounds **43** and **44** displayed a remarkable activity in both assays, while the dimethoxy compound **45** showed the best

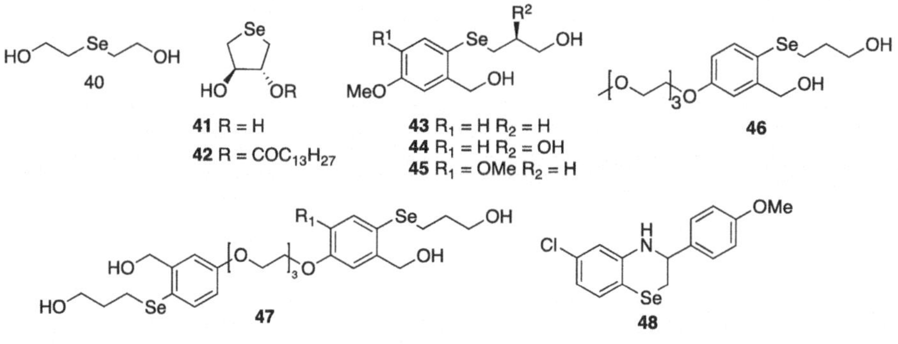

Fig. 2.8 Structures of selenides

Scheme 2.3 Catalytic cycle of selenides proposed by Back

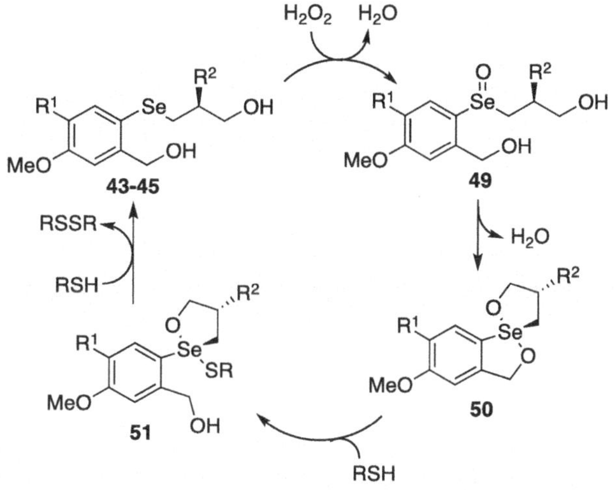

activity expressed as T_{50} (36 min measured with HPLC and 5 min with NMR). The monomer **46** and the dimer **47** displayed excellent catalytic activity and it is interesting to note that the dimeric species **47** exhibited a T_{50} roughly half that the monomer **46**, consistent with the presence of two redox centres [76]. The proposed mechanism through which these compounds exert the GPx-mimic activity is different to that reported in Scheme 2.2. As depicted in Scheme 2.3, after selenide is turned into selenoxide (**49**), the authors proposed that, because of the presence of the two properly spaced hydroxyl groups, **49** is turned into the spirodioxyselenurane **50**, which is the actual catalytic species able to convert thiols into disulfides in a two steps procedure [84].

Benzo[*b*][1,4]selenazines, like **48**, were recently prepared through an intriguing example of hetero Diels–Alder reaction and tested by the NMR-based method using DTT as thiol probe. The activity of this class of compounds is heavily influenced by the nucleophilic properties of selenium, but it is lower than that of diphenyl diselenide, which was tested in parallel [138].

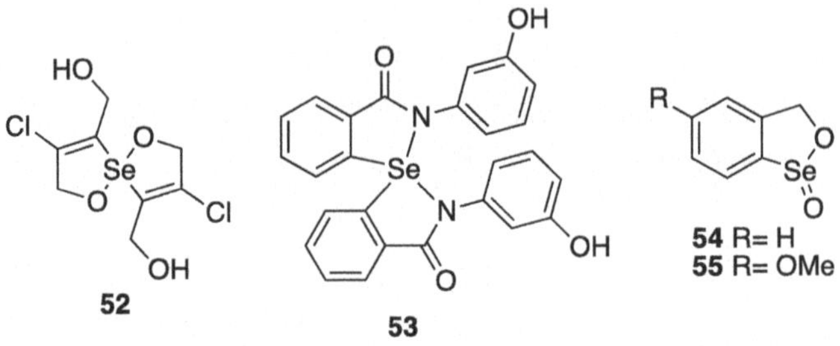

Fig. 2.9 Structures of compounds **52–55**

2.2.4 Other Selenocompounds

Starting from the assumption that the catalytic cycle of certain selenides entails the formation of spiroselenuranes intermediates (see compound **50** in Scheme 2.3 for an example), some researchers sought to develop compounds bearing selenium comprised in such functional group. Although in some cases the spirodioxoselenuranes were devoid of any GPx-mimic activity [139], in some others, as in the case of compounds **52** (Fig. 2.9), a residual pseudo enzymatic activity was recorded through a NMR-based assay [140]. Better results were obtained with spirodiazaselenuranes, indeed compound **53** has a GPx-mimic comparable to that of the corresponding selenide. From a structural point of view, selenium lays on the center of a distorted trigonal bipyramid core with two nitrogen atoms staying in the apical positions and two carbon atoms with the selenium lone pair occupying the equatorial ones. The particular structural arrangement may account for the poor GPx-mimic activity displayed by compound **53** and analogues. The mechanism of thiol oxidation should entail a redox shuttle process between the selenides and spirodiazaselenuranes via the corresponding selenoxides, similar to that reported in Scheme 2.3 [141].

Seleninate esters showed a strong GPx-like catalytic activity even if compound **54** proved to catalyze not only the oxidation of thiols to disulfides but also to thiolsulfinates, which could be detrimental in vivo in the case the native proteins and peptides are similarly affected [142]. The GPx-mimic activity can be easily increased by the addition of a methoxy substituent at *para* position to the selenium, as in compound **55** [143]. A density functional theory (DFT)-based approach was undertaken to explain why the introduction of the methoxy substituent in that particular position was so fruitful for the GPx-mimic activity. It was found that such substitution affects the rate of H_2O_2 scavenging differently, depending on the position in the aromatic ring. Activity is enhanced with methoxy in *para* position, unaffected in the *meta*, and decreased in the *ortho* one [144].

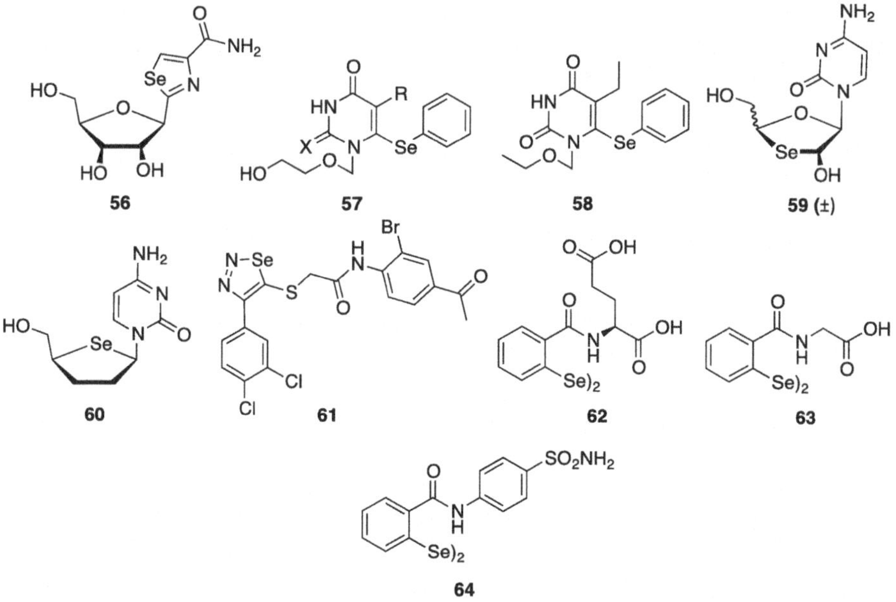

Fig. 2.10 Selenazofurin **56** and anti-HIV organoselenium compounds

2.3 Selenium-Based Antivirals

The first organoselenium compound reported for its antiviral properties was selenazofurin (compound **56**, Fig. 2.10), reported for the first time in 1983 [35]. The compound was developed as a ribavirin analogue, since ribavirin has a broad spectrum of antiviral activities both in vitro and in vivo and nowadays is used for the treatment of hepatitis C virus (HCV) and other virus associated-diseases [145]. As the parent compound, **56** exhibited a broad-spectrum antiviral activity against both DNA and RNA viruses. Of the DNA viruses, selenazofurin was able to inhibit the families Poxviridae and Herpesviridae (HSV-1, HSV-2), but the greatest antiviral activity was observed against RNA viruses, particularly toward the families Paramyxoviridae, Reoviridae, Togaviridae, and Arenaviridae. The antiviral activity of compound **56** proved to be both virucidal and virustatic, depending on the virus and cell line used to the evaluation [35]. The combination of ribavirin/compound **56** was assayed and a synergistic activity against several viruses was found [146]. As for ribavirin, selenazofurin is thought to inhibit inosine-5′-monophosphate dehydrogenase (IMPD) once converted into the corresponding nucleotide triphosphate [147]. IMDP catalyzes the rate-limiting step of guanine nucleotide metabolism, thus controlling the size of the guanine nucleotide pool, which in turn controls many physiological processes, including replication, transcription, signalling, and glycosylation [148]. In 1986, it was demonstrated that selenazofurin inhibits replication of influenza A and B viruses in vitro and that the antiviral potency was greater than that of the ribavirin [149]. Unfortunately, this result was not confirmed in vivo, where ribavirin showed a better antiviral activity [150]. Compound **56** was later tested and investigated as antitumor agent and several analogues were developed using its structure as a template [151]. Finally, in 2002 it was found that the compound is active against the west Nile virus [152].

After selenazofurin, several compounds were developed for the treatment of specific viral infections, here just the most recent examples will be detailed, according to the pathogen they were meant to inhibit.

2.3.1 Anti-HIV Compounds

The human immunodeficiency virus (HIV) is one of the most serious health threat worldwide. According to UNAIDS factsheet, there are more than 37 million HIV-positive individuals and each year roughly two million new infections are recorded [153]. Starting from 1984, when HIV was identified, unprecedented success has been achieved in discovering drugs, as reflected by the fact that there are now more drugs approved for the treatment of HIV than for all other viral infection taken together. Despite the number of commercially available drugs, there is no concrete perspective of eradication on the horizon, and this is the main reason why the anti-HIV research is yet strongly pursued. There are two main types of HIV: HIV-1, which is the most common and HIV-2. Relatively uncommon, predominantly found in West Africa, HIV-2 is less infectious, but less sensitive to some drugs.

The first report on organoselenium molecules endowed with anti-HIV properties dates back to 1991, when some acyclic selenium-containing nucleosides, related to acyclovir, were reported [154]. This design strategy was strengthened since it was simultaneously discovered that reduced levels of selenium and glutathione peroxidase activity in blood occurs in patients with AIDS [155]. Compounds **57** were potent against HIV-1 with the median effective concentration (EC_{50}) ranging from 0.96 to 13.0 μM. When tested in human peripheral blood mononuclear cells (PBMCs) infected with HIV-2, these compounds were found to have a similar activity to that exhibited against HIV-1. Surprisingly, none of the compounds was able to inhibit reverse transcriptase (RT) activity, even if their structural arrangement may suggest a similarity to nucleoside reverse transcriptase inhibitors (NRTIs) [154]. Noteworthy, by removing the hydroxyl group on the side chain, compound **58** and analogues were prepared, demonstrating that the anti-HIV-1 and HIV-2 activity of acyclic 6-phenylselenenyl pyrimidines was retained or, in certain cases, even improved [156]. Compound **58** demonstrated the striking anti-HIV-1 activity at 0.017 μM, which was however 850 times higher against HIV-2. Surprisingly, compound **58** was able to inhibit RT even if at high concentrations, suggesting that RT is not the primary target for its antiviral activity. As a consequence of the RT inhibition, it was proved that HIV-1 rapidly became resistant to compound **58** after just three passages [157]. In parallel, the compound was investigated for its pharmacokinetic features and resulted nontoxic in mice and rats [158]. The same authors developed a series of oxaselenolane nucleosides capable not only to inhibit HIV-1, but also hepatitis B virus (HBV). The cytosine analogue **59** was found to exhibit the most potent anti-HIV (EC_{50} 2.69 μM) and anti-HBV activities (EC_{50} 1.2 μM) with no in vitro toxicities up to 100 μM in various cell lines. The separation of enantiomers clearly evidenced that the (−)-isomer is more potent than the (+)-isomer [159, 160]. Inspired by the potent anti-HIV activity of 2′,3′-dideoxynucleosides (ddNs), their bioisosteric analogues 2′,3′-dideoxy-4′-selenonucleosides (**60**) were prepared, but although X-ray crystallographic analysis indicated that 4′-seleno-ddNs adopted the same C2′-endo/C3′-exo conformation as the anti-HIV active ddNs, they did not show any antiviral properties [161]. The reason behind the lack of activity is that cellular kinases are not able to convert selenonucleoside in the active nucleotide.

Se-containing non-nucleoside reverse transcriptase inhibitors (NNRTIs) were developed in 2009, one example is the selenodiazole **61**. This compound was able to inhibit HIV-1 replication at low micromolar concentrations, while showing a more pronounced cytotoxic effect if compared with the corresponding thiadiazole. The inability of **61** to inhibit HIV-2 together with the striking similarity with known NNRTIs, led the authors to classify selenodiazole as a novel class of NNRTIs [162].

Considering that HIV-1 gene expression and viral replication can be induced by oxidative stress, various antioxidants have been tested for their inhibitory effects on HIV-1 activation and were found to suppress viral replication. Among these, ebselen **1** selectively inhibited HIV-1 replication in MOLT-4 cells at a concentration of 5–10 μM [163]. The mechanism by which it exerts antiviral activity was found to be the inhibition of capsid assembly [164].

The most recent example of selenium-containing anti-HIV agents was reported by some of us as a novel class of nucleocapsid protein 7 (NCp7) inhibitors, called diselenobisbenzamides (DISeBAs) [165]. NCp7 is a small (55 amino acids), highly basic protein, which plays pivotal roles within the HIV replicative cycle. It is highly conserved in viral strains and its inhibition did not select for resistant strains. From a structural point of view, the protein is characterized by the presence of two zinc finger (ZF) motifs, each comprising 14 amino acids, organized in a C-X2-C-X4-H-X4-C pattern. Several compounds acting in the ZF domain were reported and they are classified as covalent (also known as zinc ejectors), non-covalent and nucleic acid binders NCp7 inhibitors [166, 167]. The anti-HIV diselenides (**62**–**64**) belong to the class of covalent inhibitors because they share striking similarities to known zinc ejectors and are thus thought to remove the zinc ions from the ZF domains. Compound **62** is the prototypical of amino acid amides, the series that showed the best biological profile, being endowed with potent and selective anti-HIV activity. The aromatic amides, such as **64**, has similar anti-HIV potency, but coupled with a pronounced toxicity, thus resulting in lower selectivity indexes. Time of addition experiments indicated that compound **62** interfered with early entry, whereas early and late NCp7 functions were the target of the glycine (**63**) and isoleucine derivatives, respectively. Compound **63** was able to exert a strong virucidal effect, although at higher concentrations compared with other known NCp7 inhibitors. The antiviral evaluation was extended to chronically infected cells, as well as to resistant HIV-1 strains, and the organoselenium compounds retained their broad-spectrum antiviral

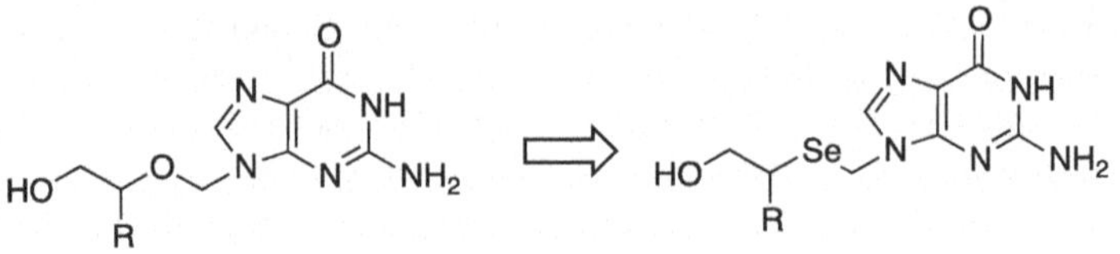

65 R = H **acyclovir**
66 R = CH_2OH ganciclovir

67 R = H seleno-acyclovir
68 R = CH_2OH seleno-ganciclovir

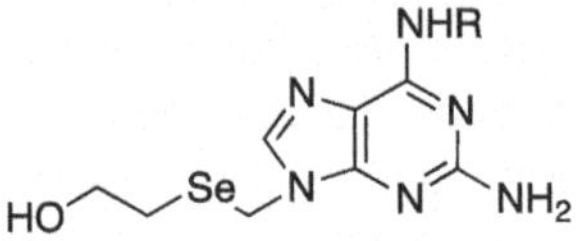

69 R = Me r
70 R = H

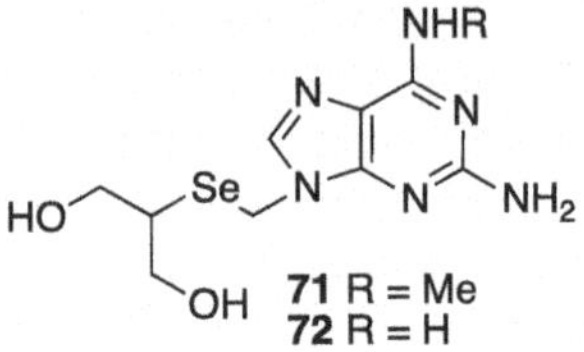

71 R = Me
72 R = H
73 R = 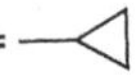

Fig. 2.11 Organoselenium compounds with anti-herpetic activity

properties, which comprises also clinical isolates and HIV-2. The proteomic analysis of DISeBA-treated latently infected cells revealed the accumulation of unprocessed Gag polyprotein, indicating that DISeBAs were able to recognize NCp7 even in the context of the Gag precursor. Noteworthy, molecular dynamic simulations shed a light on the DISeBAs/NCp7 reaction [165].

2.3.2 *Anti-herpes Virus Compounds*

Herpesviridae is a large family of DNA viruses able to produce latent and recurring infections in animals, including humans. Nowadays, there are nine herpes virus types known to infect humans: herpes simplex viruses 1 and 2 referred to as HSV-1 and HSV-2, varicella-zoster virus (VZV), Epstein–Barr virus (EBV), human cytomegalovirus (HCMV), human herpes virus 6A and 6B, human herpes virus 7, and Kaposi's sarcoma-associated herpes virus (KSHV) [168]. The most commonly used drugs against herpetic viruses are acyclovir (**65**, Fig. 2.11) and ganciclovir (**66**). Acyclovir, which is the prototype of acyclic nucleosides, is converted into the corresponding nucleotide triphosphate through a series of three sequential phosphorylation reactions, the first of which is carried out by the viral encoded thymidine kinase. The thymidine kinase-mediated activation endows the drug with a high selectivity toward infected cells [169]. Ganciclovir **66** has been clinically developed against HCMV infection. In HCMV infected cells, it is converted to the monophosphate by a phosphotransferase and then subsequently converted to the diphosphate by cellular guanylate kinase and into the triphosphate by a number of cellular enzymes. Ganciclovir triphosphate competitively inhibits HCMV polymerase by reacting with the cellular deoxyguanosine triphosphate. Very recently, the corresponding selenium-containing analogues **67** and **68** were developed and

Table 2.3 Anti-herpesvirus activity of organoselenium compounds **55–63**

Compound	EC_{50} (μM)				Ref
	HSV-1	HSV-2	VSV	HCMV	
55	0.66	1.02	6.4	18.9	[170]
56	0.90	1.40	11.1	2.14	[170]
57	1.47	6.34	>100	>100	[170]
58	>100	>100	>100	53.1	[170]
59	14.3	17.6	>100	>100	[171]
60	15.4	23.2	>100	>100	[171]
61	>100	>100	>100	32.1	[171]
62	>100	>100	>100	34.3	[171]
63	>100	>100	>100	41.1	[171]

assayed for the antiviral activity against HSV-1, HSV-2, VZV, and HCMV, with their cytotoxicity being parallelly tested. Both compounds did not show any toxic effect up to the highest tested concentration of 100 μM. Seleno-acyclovir **67** showed potent antiviral activity only against HSV-1 and HSV-2, whereas compound **68** exerted moderate activity only against HCMV. However, **67** and **68** were found to be inactive against VZV [170]. Successively, a SAR study was carried out by modifying the purine scaffold and the effect of such modifications on the antiviral activity was assayed (compounds **69–73**). As reported in Table 2.3, compound **71** exhibited the most potent anti-HCMV activity, which was higher than the hit compound **68**. Compounds **69** and **70** showed significant antiviral activities against HSV-1 and HSV-2, indicating that they might be deaminated by the cellular nucleoside deaminase to serve as prodrugs of seleno-acyclovir (**67**). Compounds **71–73**, exhibited significant anti-HCMV activity, which was higher than that showed by compound **68**. As selenol-acyclovir and selenol-ganciclovir, compounds **71–73** were completely inactive toward VZV. From a general point of view, it is important to stress that all of the compounds were less potent than the reference ones **65** and **66**, probably because of the difficulty in phosphorylation, caused by steric effects of the bulky selenium atom [171, 172].

In 2016, it was demonstrated the anti-HSV-2 action of diphenyl diselenide **19** in vitro and in vivo. Diphenyl diselenide showed an antiviral action against infected Vero cells culture and reduced the histological damage, extravaginal lesion scores, and the viral load of vaginal tissue of mice. Such antiviral activity was related to the immunomodulatory, antioxidant, and anti-inflammatory properties of the compound. A virucidal effect was supposed to be responsible for the antiviral action because a direct effect on the viral particles was observed since the infectivity of HSV-2 was sensibly reduced upon pretreatment with compound **19**. The mice infected with HSV-2 showed signals of extravaginal lesion, that were reduced by the administration of **19** to the same extent to that showed by acyclovir [173]. Successively, it was demonstrated that the treatment with the ebselen-derivative **12** minimized the hepatic and renal oxidative stress and the toxicity caused by HSV-2 infection in mice. In general, the effects of the isoselenazolone **12** were similar to those exerted by acyclovir, but different for what concern the renal adenosine deaminase and hepatic catalase activities, enzymes thought to be associated to renal and hepatic damage in the frame of viral infections [174].

2.3.3 Virucidal Compounds and Ebselen as Anti-HCV Agents

HCV infection, with an estimated 130–170 million chronically infected individuals, surely represents a serious health problem worldwide. It is capable to cause serious liver diseases, such as cirrhosis, finally leading to hepatocellular carcinoma [175]. After the identification of the virus, tremendous efforts have been pushed toward the development of agents able to inhibit viral proteins. These efforts led to the design of many antivirals, most of which target the non-structural protein 3 (NS3) with protease activity, the NS5B polymerase, or the NS5A RNA binding protein. Three of these NS3 protease inhibitors and one NS5B polymerase inhibitor have been recently approved by FDA to treat HCV [176]. Among the HCV encoded proteins, the NS3 protein is the only known enzyme that contains both protease and helicase active sites. The helicase catalytic activity is remarkably difficult to inhibit with small molecules. Several high-throughput screens strategies identified few inhibitors, and most of them are either toxic or do not act as antivirals in a cell context. In 2014, the NCC was screened against this difficult target identifying only one specific helicase inhibitor, ebselen **1** [41]. By the modification of key cysteine residues, it was indeed able to selectively block the helicase function of NS3 without affecting its protease activity. Actually, the anti-HCV activity of ebselen was reported 4 years before, in 2010, even if the molecular target, at that time, was unknown [177, 178].

In a SAR study that lasts more than a decade, the virucidal activity of a series of selenium-containing molecules was investigated. It worth to stress out that the virucidal is quite different from the antiviral activity and frequently happens that a virucidal compound does not function as antiviral and vice versa. Compounds **74–86** [179–184] (Fig. 2.12) were tested against encephalomyocarditis virus (EMCV), vesicular stomatitis virus (VSV), and HSV-1.

ECMV is a small non-enveloped single-strand RNA virus, that is the causative agent of myocarditis, encephalitis, neurological diseases, reproductive disorders and diabetes in many mammalian species [185]. VSV is an enveloped RNA virus responsible of vesicular stomatitis in humans and livestock [186].

As reported in Table 2.4, all the compounds were tested in parallel for their cytotoxicity in the same cell line used for viral titration. Noteworthy, ebselen was used as the positive control compound. Among the tested compounds, just a limited number was able to exert virucidal activity against VSV, the best compound was the aza analogue of ebselen **76**, lacking the *N*-phenyl ring, which is also the most potent against HSV-1 [180]. The most potent ECMV

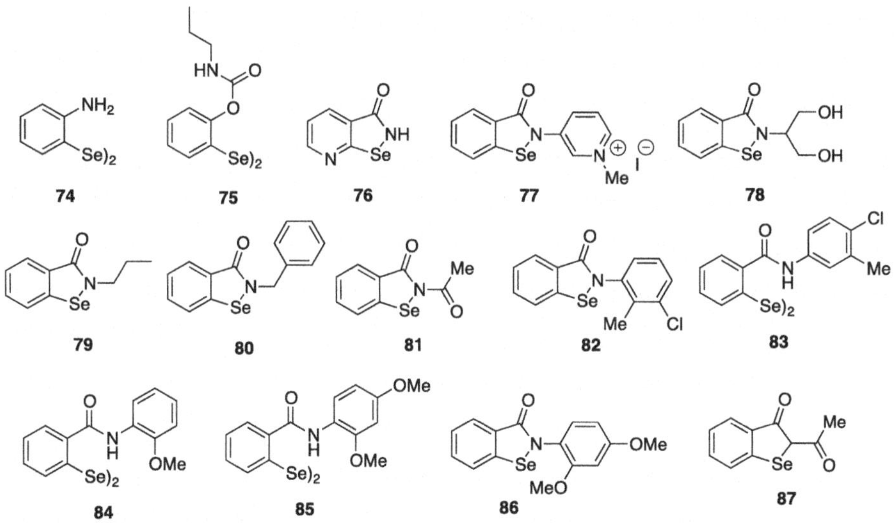

Fig. 2.12 Compounds endowed with virucidal activity

Table 2.4 Virucidal activity of organoselenium compounds **74**–**84**

Compound	MIC (μg/ml)				
	HSV-1	ECMV	VSV	CC_{50}	Ref
74	40	10	>1000	7	[179]
75	>1000	4	400	15	[179]
76	0.4	6	80	25	[180]
77	20	20	>1000	80	[180]
78	4	8	>1000	15	[181]
79	6	6	>1000	–	[182]
80	2	6	>1000	–	[182]
81	4	6	600	–	[182]
82	6	–	–	78	[183]
83	10	–	–	58.5	[183]
84	6	400	>1000	31.2	[184]
85	2	600	>1000	58	[184]
86	2	200	>1000	156	[184]
Ebselen **1**	2	10	>1000	15	[179]

virucidal compound was diselenide **75**, which, most probably is able to transfer the carboamoyl group to the viral surface proteins, thus blocking virus entrance into the cell [179]. The virucidal activity of the ebselen-like compound **81** is due to the precise arrangement of the selenoamidic group. Indeed, the sole presence of selenium, as in compound **87**, is not enough to obtain a virucidal activity [182].

2.4 Selenium-Based Antibacterial Agents

Bacterial infections can be considered a serious problem even if the availability of antibiotic and antibacterial agents partially attenuated the morbidity and the mortality caused by some infections. In the present and in the future times, the real challenge the scientific community has to deal with is the onset of multi-drug resistant bacterial strains. Indeed, new resistance mechanisms are emerging globally, sensibly reducing our ability to treat common infections, thus provoking prolonged illness, disability, and sometimes death [187]. Biofilms are among the most resistance facilitating elements known to date, they are microbial sessile communities in which microorganisms live stacked each other, in a matrix composed of proteins, lipids, and polysaccharides. In biofilms, microorganisms are more resistant to antimicrobial drugs and immune system responses with respect to the free or planktonic form [188]. Considering that it is impossible to list all of the examples of organoselenium compounds endowed with antibacterial properties, a selection will be mainly made on the basis of the novelty while a special attention will be devoted to those compounds displaying activity against *Staphylococcus aureus*, as representative of gram-positive bacteria and *Escherichia coli*, as representative of gram-negative ones. Anti-biofilm compounds will be also listed together with information about the mechanism of action by which these compounds exert the antibacterial activity.

2.4.1 Ebselen and Ebselen-Like Structures

Ebselen **1** was first profiled for its antibacterial spectrum in 1989, in particular it was found active against methicillin-resistant *S. aureus* (MRSA) strains [189]. MRSA was first isolated in 1961 and its increasing prevalence poses serious clinical and epidemiological issues all over the world [190]. In 2013, the Centers for Disease Control and Prevention (CDC) reported that more than 11,000 people died from a MRSA-related infection, representing nearly half of all fatalities caused by antibiotic-resistant bacteria [191]. In 2015, the antimicrobial activity of ebselen was tested against a panel of clinical isolates of multi-drug resistant *S. aureus*, and showed potent bactericidal activity with MICs ranging from 0.125 to 0.5 μg/ml against not only MRSA, but also vancomycin-resistant and linezolid-resistant (VRSA) strains. The antibacterial properties of compound **1** translated into a potent ability to reduce adherent biofilms of *S. aureus* and *S. epidermidis*. Then, ebselen was tested in vivo in a mouse model of MRSA skin infection. One and 2% Ebselen in petroleum jelly significantly reduced the mean bacterial counts compared to the control group. Inhibition of protein synthesis at a concentration equivalent to the MIC demonstrates that protein synthesis is likely the primary antibacterial mechanism of action of compound **1**. A synergistic activity with topical antimicrobials was finally observed, providing a strong rationale for the clinical usage of this organoselenium compound [192]. Among gram-positive bacteria, besides MSRA, ebselen **1** showed potent activity against clinical isolates of *Enterococcus faecalis* and *E. faecium*, with MIC_{90} of 0.5 μg/ml. It also showed a potent activity against vancomycin-resistant strains of *Enterococcus*. On the other hand, compound **1** did not exert antimicrobial activity against gram-negative pathogens. Rather than lack of target, this can be explained by a reduced permeability to the outer membrane barrier or to the efflux pumps, which move the compound outside the cell [193].

A further application of ebselen is in the treatment of *Clostridium difficile* infection (CDI). In particular, it was proved that, targeting the cysteine protease domain (CPD) within the *C. difficile* major virulence factor toxin B (TcdB), ebselen blocked *C. difficile* growth with micromolar potency. TcdB is composed of a putative receptor binding domain, a transmembrane domain, a CPD, and a glucosyltransferase domain (GTD). When endocytosed by host cells, the bacterial toxin exposes the CPD to the mammalian-specific cytosolic sugar 1D-myo-inositol hexakisphosphate (IP6). The allosteric binding of IP6 activates the CPD to autocatalytically cleave the GTD domain, which induces toxicity by glucosylation of the GTPases in host intestinal epithelial cells. This event results in rearrangement of the actin cytoskeleton, acute inflammation, massive fluid secretion, and finally necrosis of the mucosal layer of the colon. Ebselen was able to block the toxin by covalently modifying the CPD, which takes place after the allosteric modification mediated by IP6. Treatment in a mouse model of CDI, that closely resembles the human infection, confirmed the therapeutic benefit of ebselen [194].

The antibacterial spectrum of ebselen comprises also *Mycobacterium tuberculosis*, which is a tremendous health problem worldwide because of its ability to turn into mutant strains. In the recent years, several *M. tuberculosis* infections were multidrug-resistant, which appear to be untreatable with current antitubercular therapies. Quite recently, the mycobacterium-encoded antigen 85 (Ag85) complex was recognized as a viable target because of its importance in the synthesis of mycolic acids, that are essential in the structure of the *M. tuberculosis* cell wall. In a target-based drug discovery campaign, ebselen was found as a covalent Ag85 inhibitor, in particular mass spectrometry experiments conclusively established that it was able to react with Cys209, making a stable selenylsulfide intermediate able to block the protein activities [195].

In the frame of antibacterial agents, the ability of ebselen to modify key cysteines was proved for the inhibition of New Delhi metallo-β-lactamase (NDM-1). NDM-1 is one of the most important metallo-β-lactamase responsible for the resistance to first and second lines β-lactams, such as carbapenems in gram-negative bacteria. Ebselen was found to bind C221 of the NDM-1 of MDR *E. coli*, restoring the activity of meropenem [196]. Besides NMD-1, ebselen reacts with the dithiol Cys135-Cys138 present in the *E. coli* TrxR active site, with a dissociation constant of 0.14 μM [197].

Bacterial urease is a key enzyme that facilitates the infectivity by modifying the pH of the target tissue. This enzyme is the principal virulence factor of several microorganisms, such as *Helicobacter pylori*. Ebselen was found to inhibit urease

Fig. 2.13 Ebselen-like compounds as antibacterial agents

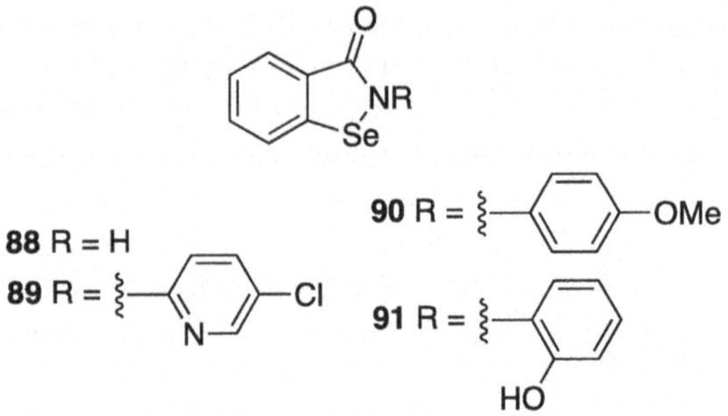

Table 2.5 Antibacterial activity of compounds **77**, **78**, **88–91**

Compound	MIC (μg/ml)		Ref
	S. aureus	*E. coli*	
77	8[a]	16[b]	[180]
78	2[c]	14[d]	[181]
88	6[c]	25[d]	[181]
89	2[e]	–	[199]
90	10[a]	152[b]	[183]
91	8[f]	32[g]	[184]

[a]Tested on PCM1944 strain
[b]Tested on PCM2057 strain
[c]Tested on ATC25923 strain
[d]Tested on ATC25922 strain
[e]Tested on ATC29213 strain
[f]Tested on ATCC6538 strain
[g]Tested on ATCC10563 strain

with a Ki in the nanomolar range by covalently modifying Cys322 that is positioned in the narrow active site of the enzyme [198].

During the years, several ebselen-like structures have been prepared and tested against several pathogenic bacteria, here a selection (Fig. 2.13) will be reported.

As reported in Table 2.5, the compounds showed a potent anti-bacterial activity, which is more pronounced against *S. aureus* in comparison to *E. coli*. This can be plausibly explained by the lack of cellular penetration in the gram-negative bacteria; however, it is impossible to draw any comparison since these activities were taken from articles where the compounds were assayed against different bacterial strains.

2.4.2 *Diselenides, Selenides, and Selenones*

Diselenides **74** and **75** (Fig. 2.12) were reported for their antibacterial activities even if they exerted a lower potency if compared with ebselen [179]. While the amphetamine-derived diselenide **92** (Fig. 2.14) was tested in the frame of a wider

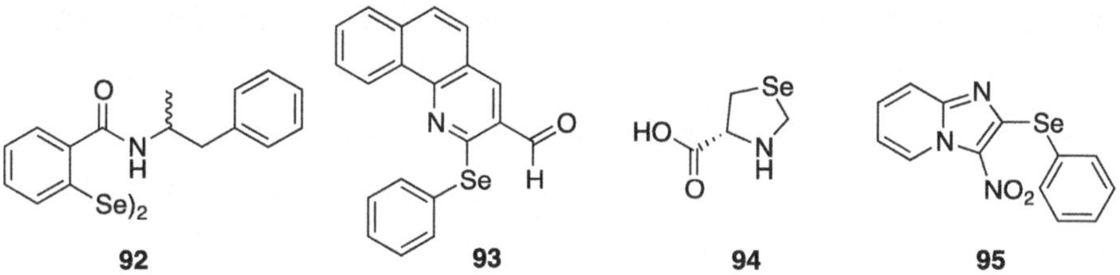

92 93 94 95

Fig. 2.14 Antibacterial selenocompounds

Table 2.6 Antibacterial activity of compounds **74**, **75** and **92–95**

Compound	MIC (µg/ml)		Ref
	S. aureus	*E. coli*	
74	17.3[a]	173[b]	[179]
75	17.2[a]	173[b]	
92	31.25[c]	–	[200]
93	8[d]	25[e]	[201]
94	–	2.5[f,g]	
95		2.5[h]	

[a]Tested on CCM3953 strain
[b]Tested on CCM398 strain
[c]Tested on ATCC 29213 strain
[d]Tested on ATC29737 strain
[e]Tested on ATC20852 strain
[f]Tested on ATC25922 strain
[g]Data in µM
[h]Tested on MTCC2961 strain

study, aimed to the identification of the structural characteristics around the diselenide moiety to obtain antibacterial compounds. Compound **92** showed a moderate antibacterial activity against *S. aur*eus (Table 2.6), but demonstrated considerable antibacterial activity against *S. epidermidis* and *S. pyogenes*. When tested at a concentration lower than the MICs, it proved to reduce the formation of biofilms of the gram-positive bacteria *S. pyogenes*. In addition, it was able to dissolve a preformed biofilm of *S. epidermidis* but unfortunately, it was quite toxic when tested in Hela cell [200].

Among selenides, derivative **93** was deeply profiled for its antibacterial activity against gram-positive and gram-negative strains. In particular, besides the MICs (Table 2.6) on six bacterial strains, the wound healing abilities were ascertained using an in vivo model. Most probably the biological activities are linked to the compound ability to bind DNA and to its antioxidant property, the latter measured by different methods [201]. L-Selenaproline, compound **94**, was the unique proline analogue able to inhibit *E. coli* growth in Mueller–Hinton medium and in human urine, as well as in glucose minimal medium. Interestingly, compound **94** exerted antibacterial activity under conditions that may occur in the urinary tract [202]. In 2014, it was proved that the selenaproline, as well as selenocystine **4** (Fig. 2.4), are internalized by the cysteine transport system in *E. coli* [203]. The imidazo[1,2-*a*] pyridine **95** showed a pronounced anti *E. coli* activity, quite similar to that of rifampicin, which was used as positive control. Interestingly, compound **95** showed a synergistic effect not only with kanamycin, but also with rifampicin [204].

2.5 Selenium-Based Antifungal and Antiprotozoal Compounds

Fungi can be pathogenic microorganisms for humans and, upon treatment, they may show acquired drug resistance. The past two decades have seen an increase in the incidence of life-threatening infections, largely due to the extensive use of broad-spectrum antibacterial agents, that facilitate opportunistic fungal development. In addition, the number of patients with immunosuppression due to AIDS or anticancer treatment regimens are increasing, thus enhancing the likelihood of infection by fungi otherwise well tolerated. Compared with antibiotics, a lower number of drugs are currently approved to treat fungal infections highlighting the need for the development of antifungal compounds. An important group of opportunistic fungal pathogens is represented by the genus *Candida*, which includes more than 200 species of yeasts. These yeasts are a common cause of hospital-acquired bloodstream infections, since they are able to stick on inert surfaces such as the resin of artificial dentures and catheters. As a result, these devices surely represent a risk source for highly invasive candidiasis, which is one among the most common systemic fungal infection [205].

Ebselen was reported to be endowed with anti-*Candida* activity in 2004 [180], due to its ability to inhibit the plasma membrane H^+-ATPase [206], which is essential to establish proton gradients across the plasma membrane and to maintain a proper intracellular pH. In 2009, ebselen was tested and proved to inhibit the growth of two strains of fluconazole-resistant *C. albicans* (S2 and ATCC 96901) in a concentration dependent manner. The ebselen fungicidal/fungistatic effect is concentration dependent, at 10 μM is thought to be fungistatic, while at 30 μM it is fungicidal [207].

Diphenyl diselenide **19** is the most studied organoselenium compound from the antifungal activity point of view. It was tested against 44 strains of yeasts and seven strains of molds. It inhibited *C. albicans* growth with a minimal inhibitory concentration (MIC) of 8.35 μg/ml, and it exhibited a fungicidal action, expressed as minimal fungicidal concentration (MFC), of 11.81 μg/ml [208]. Two further derivatives (compounds **96** and **97**, Fig. 2.15) were also tested, but they displayed a lower activity when compared with the lead compound **19**. The antifungal activity was later ascribed to the ability of **19** to irreversibly damage *C. albicans* cells. This is quite interesting, since it was proved that diphenyl diselenide is able to trigger oxidative stress in *Saccharomyces cerevisiae*. Thus, the mechanism responsible for the antifungal activity depends on the pathogen the compound is tested against.

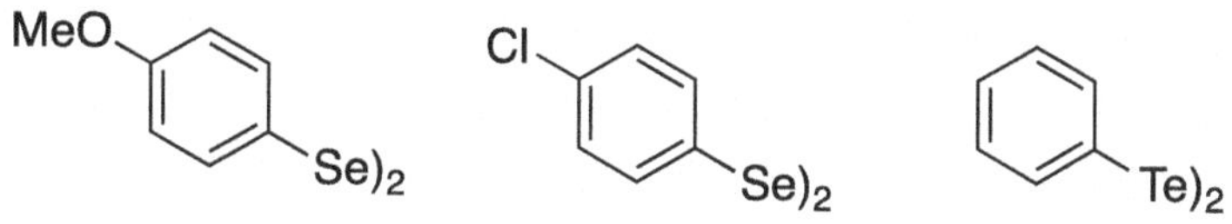

96 **97** **98**

Fig. 2.15 Diphenyl diselenide analogues with antifungal activity

Interestingly, the concentration that inhibited *C. albicans* growth is within the plasma concentrations observed when the compound was administered to rodents. The telluro-analogue of diphenyl diselenide **98** showed a comparable anti-*Candida* activity [209].

Among Candida species, *C. glabrata* has emerged as the cause of candidiasis and it is less susceptible to azole antifungals, such as fluconazole, as well as to amphotericin B. In addition, *C. glabrata* systemic infection has been associated with high mortality, especially in patients with weakened immune system. In 2013, diphenyl diselenide proved to inhibit *C. glabrata* growth showing a synergistic activity with amphotericin B [210].

Protozoa are unicellular, eukaryotic organisms, some of whom responsible of severe diseases, such as malaria, African sleeping sickness (ASS), and leishmaniasis. Malaria is a global health concern, infecting more than 200 million people each year and it is caused by a group of protozoa called *Plasmodium vivax*, *P. ovale*, *P. malariae*, and *P. falciparum*. *P. falciparum* is responsible for most of the annual mortality, because of its ability to establish a cerebral infection [211]. In the Plasmodium life cycle, glucose metabolism is very important and molecules able to selectively alter protozoan glycolysis are thought to be good candidates for the malaria treatment. Among the targets, the *P. falciparum* Hexokinase (PfHK), the enzyme which converts glucose into glucose-6-phosphate, shares limited structural similarity (24%) with human glucokinases (HsGlk, or HK IV) and is considered as a suitable target for the development of innovative therapeutics. Very recently, ebselen was identified as a very potent inhibitor of PfHK with a IC_{50} of 0.01 μM. Interestingly, the ebselen-mediated inhibition is not related to cysteine modifications, as observed by the reversal of the enzyme activity after sequential dilution. The ability to inhibit the enzyme turned into a good toxicity to *P. falciparum* with EC_{50} of 6.8 μM [212]. This value is in agreement with that found in 1989, when the antimalarial activity of ebselen was reported for the first time without any indication regarding the mechanism of action [213].

The Hexokinase was used as target to discover compounds able to cure African sleeping sickness, which is caused by the bloodstream form of *Trypanosoma brucei* and transmitted by the tsetse fly [214]. *T. b. gambiense* and *T. b. rhodesiense*, responsible for the more acute East African sleeping sickness (ASS), are two subspecies of *T. brucei* able to cause human disease. Although treatments for ASS are available, their efficacy is compromised by serious adverse side effects, such as acute toxicity, neural disorders, and in some cases death. Among the alternative, unexploited

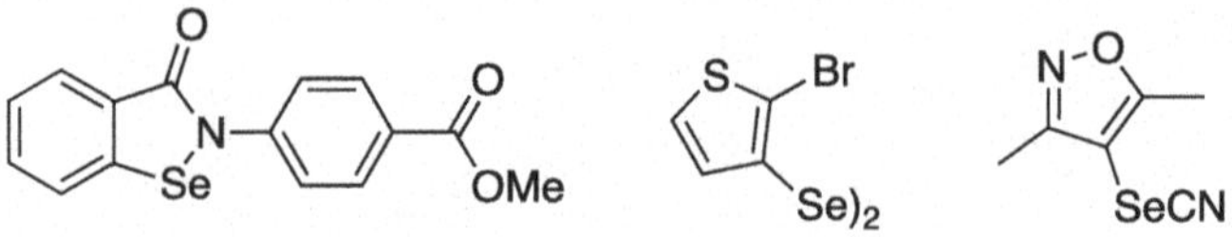

Fig. 2.16 Antiprotozoan compounds

options, TbHK1 is a well-validated target, which was subjected to high-throughput screening efforts leading to the identification of ebselen. The organoselenium compound was able to inhibit the enzyme with an IC_{50} of 0.05 μM. Notably, the inhibition was, this time, irreversible [215]. In a follow-up study, compound **99** (Fig. 2.16) was identified as the only molecule able to potently inhibit TbHK1 and exert trypanocidal activity [216].

Leishmaniasis is an infectious poverty-associated disease caused by protozoan parasites of the genus *Leishmania*. Leishmaniasis are vector-borne diseases with great epidemiological and clinical diversity. Three clinical types of leishmaniasis are known: cutaneous, mucocutaneous, and visceral (also known as kala-azar), which differ in their immunopathology and degrees of morbidity and mortality [217]. Starting from the assumption that the increment of selenium level in blood is a defensive strategy the body undertakes upon *Leishmania* infection [218], Sanmartin and coworkers designed a large series of diselenides and selenocynates endowed with anti-*Leishmania* properties through the inhibition of the key protozoan enzyme trypanothione reductase (TryR) [219]. TryR catalyzes the reduction of trypanothione disulfide to trypanothione. From a general point of view, diselenides are more potent and selective than the corresponding selenocyanate derivatives. Among the whole set of tested compounds, diselenide **100** and selenocyanates **101** are worth mentioning because of their potent anti-*Leishmania* activity, which properly correlated with the inhibition of TryR [219].

References

1. Reilly C (2006) Selenium in food and health. Springer Science + Business Media, LLC, New York
2. Shao S, Zheng B (2008) The biogeochemistry of selenium in Sunan grassland, Gansu, Northwest China, casts doubt on the belief that Marco Polo reported selenosis for the first time in history. Environ Geochem Health 30:307–314
3. Frost DLV (1965) Selenium and poultry. Worlds Poult Sci J 21:139–156
4. Gnadinger CB (1933) Selenium insecticide material for controlling red spider. Ind Eng Chem 25:633–637
5. Nelson EM, Hurd-Karrer AM, Robinson WO (1933) Selenium as an insecticide. Science 78:124
6. Nelson AA, Fitzhugh OG, Calvery HO (1943) Liver tumors following cirrhosis caused by selenium in rats. Cancer Res 3:230–236

7. Frost DLV, Olson OE (1972) The two faces of selenium - can selenophobia be cured? CRC Crit Rev Toxicol 1:467–514
8. Casey CE (1988) Selenophilia. Proc Nutr Soc 47:55–62
9. Vinceti M, Crespi CM, Malagoli C, Del Giovane C, Krogh V (2013) Friend or foe? The current epidemiologic evidence on selenium and human cancer risk. J Environ Sci Health C Environ Carcinog Ecotoxicol Rev 31:305–341
10. Clark LC, Combs GF, Turnbull BW, Slate EH, Chalker DK, Chow J, Davis LS, Glover RA, Graham GF, Gross EG, Krongrad A, Lesher JL, Park HK, Sanders BB, Smith CL, Taylor JR (1996) Effects of selenium supplementation for cancer prevention in patients with carcinoma of the skin. A randomized controlled trial. Nutritional Prevention of Cancer Study Group. JAMA 276:1957–1963
11. Lippman SM, Goodman PJ, Klein EA, Parnes HL, Thompson IM, Kristal AR, Santella RM, Probstfield JL, Moinpour CM, Albanes D, Taylor PR, Minasian LM, Hoque A, Thomas SM, Crowley JJ, Gaziano JM, Stanford JL, Cook ED, Fleshner NE, Lieber MM, Walther PJ, Khuri FR, Karp DD, Schwartz GG, Ford LG, Coltman CA (2005) Designing the selenium and vitamin E cancer prevention trial (SELECT). J Natl Cancer Inst 97:94–102
12. Goossens M, Zeegers M, Van Poppel H, Joniau S, Ackaert K, Ameye F, Billiet I, Dillen K, Goeman L, Van Bruwaene S, Van der Aa F, Vekemans K, Buntinx F (2015) Phase III randomised chemoprevention study of selenium on the recurrence of non-invasive bladder cancer. The SELEnium and BLAdder Cancer Trial (SELEBLAT). Arch Public Heal 73:P5
13. Vinceti M, Solovyev N, Mandrioli J, Crespi CM, Bonvicini F, Arcolin E, Georgoulopoulou E, Michalke B (2013) Cerebrospinal fluid of newly diagnosed amyotrophic lateral sclerosis patients exhibits abnormal levels of selenium species including elevated selenite. Neurotoxicology 38:25–32
14. Rees K, Hartley L, Day C, Flowers N, Clarke A, Stranges S (2013) Selenium supplementation for the primary prevention of cardiovascular disease. Cochrane Database Syst Rev (1):CD009671
15. Rocourt CRB, Wu M, Chen BPC, Cheng WH (2013) The catalytic subunit of DNA-dependent protein kinase is downstream of ATM and feeds forward oxidative stress in the selenium-induced senescence response. J Nutr Biochem 24:781–787
16. Vinceti M, Filippini T, Cilloni S, Bargellini A, Vergoni AV, Tsatsakis A, Ferrante M (2017) Health risk assessment of environmental selenium: emerging evidence and challenges. Mol Med Rep 15:3323–3335
17. Schwarz K, Foltz CM (1957) Selenium as an integral part of factor 3 against dietary necrotic liver degeneration. J Am Chem Soc 79:3292–3293
18. Pinsent J (1954) The need for selenite and molybdate in the formation of formic dehydrogenase by members of the coli-aerogenes group of bacteria. Biochem J 57:10–16
19. Muth OH, Oldfield JE, Remmert LF, Schubert JR (1958) Effects of selenium and vitamin E on white muscle disease. Science 128:1090–1090
20. Patterson EL, Milstrey R, Stokstad EL (1975) Effect of selenium in preventing exudative diathesis in chicks. Proc Soc Exp Biol Med 95:617–620
21. Wu SH, Oldfield JE, Whanger PD, Weswig PH (1973) Effect of selenium, vitamin E, and antioxidants on testicular function in rats. Biol Reprod 8:625–629
22. Rotruck JT, Pope AL, Ganther HE, Swanson AB, Hafeman DG, Hoekstra WG (1973) Selenium: biochemical role as a component of glutathione peroxidase. Science 179:588–590
23. Flohe L, Günzler WA, Schock HH (1973) Glutathione peroxidase: a selenoenzyme. FEBS Lett 32:132–134
24. Oh SH, Ganther HE, Hoekstra WG (1974) Selenium as a component of glutathione peroxidase isolated from ovine erythrocytes. Biochemistry 13:1825–1829
25. Nakamura W, Hosoda S, Hayashi K (1974) Purification and properties of rat liver glutathione peroxidase. Biochim Biophys Acta Enzymol 358:251–261

26. Turner DC, Stadtman TC (1973) Purification of protein components of the Clostridial glycine reductase system and characterization of protein A as a selenoprotein. Arch Biochem Biophys 154:366–381
27. Andreesen JR, Ljungdahl LG (1973) Formate dehydrogenase of Clostridium thermoaceticum: incorporation of selenium-75, and the effects of selenite, molybdate, and tungstate on the enzyme. J Bacteriol 116:867–873
28. Stadtman TC (1980) Selenium-dependent enzymes. Annu Rev Biochem 49:93–110
29. Wendel A, Pilz W, Ladenstein R, Sawatzki G, Weser U (1975) Substrate-induced redox change of selenium in glutathione peroxidase studied by X-ray photoelectron spectroscopy. Biochim Biophys Acta Enzymol 377:211–215
30. Forstrom JW, Zakowski JJ, Tappel AL (1978) Identification of the catalytic site of rat liver glutathione peroxidase as selenocysteine. Biochemistry 17:2639–2644
31. Cone JE, Del Rio RM, Davis JN, Stadtman TC (1976) Chemical characterization of the selenoprotein component of Clostridial glycine reductase: identification of selenocysteine as the organoselenium moiety. Proc Natl Acad Sci U S A 73:2659–2663
32. Brigelius-Flohé R, Flohé L (2017) Selenium and redox signaling. Arch Biochem Biophys 617:48–59
33. Yang GQ, Ge KY, Chen JS, Chen XS (1988) Selenium-related endemic diseases and the daily selenium requirement of humans. World Rev Nutr Diet 55:98–152
34. Vanderpas JB, Contempré B, Duale NL, Goossens W, Bebe N, Thorpe R, Ntambue K, Dumont J, Thilly CH, Diplock AT (1990) Iodine and selenium deficiency associated with cretinism in northern Zaire. Am J Clin Nutr 52:1087–1093
35. Kirsi JJ, North JA, McKernan PA, Murray BK, Canonico PG, Huggins JW, Srivastava PC, Robins RK (1983) Broad-spectrum antiviral activity of 2-Beta-D-ribofuranosylselenazole-4-carboxamide, a new antiviral agent. Antimicrob Agents Chemother 24:353–361
36. Müller A, Cadenas E, Graf P, Sies H (1984) A novel biologically active seleno-organic compound-1 glutathione peroxidase-like activity in vitro and antioxidant capacity of PZ 51 (Ebselen). Biochem Pharmacol 33:3235–3239
37. Santi C, Marini F, Lenardão EJ (2017) Synthetic advances on bioactive selenium compounds. Chapter 2. Looking beyond the traditional idea of glutathione peroxidase mimics as antioxidants. In: organoselenium compounds in biology and medicine. Royal Society of Chemistry, Cambridge, pp 35–76
38. Pacuła AJ, Mangiavacchi F, Sancineto L, Lenardão EJ, Ścianowski J, Santi C (2016) An update on "selenium containing compounds from poison to drug candidates: a review on the GPx-like activity". Curr Chem Biol 9:97–112
39. Lesser R, Weiss R (1924) Uber Selenhaltige Aromatische Verbindungen (VI). Chem Ber 57:1077–1082
40. Singh N, Halliday AC, Thomas JM, Kuznetsova OV, Baldwin R, Woon ECY, Aley PK, Antoniadou I, Sharp T, Vasudevan SR, Churchill GC (2013) A safe lithium mimetic for bipolar disorder. Nat Commun 4:1332
41. Mukherjee S, Weiner WS, Schroeder CE, Simpson DS, Hanson AM, Sweeney NL, Marvin RK, Ndjomou J, Kolli R, Isailovic D, Schoenen FJ, Frick DN (2014) Ebselen inhibits hepatitis C virus NS3 helicase binding to nucleic acid and prevents viral replication. ACS Chem Biol 9:2393–2403
42. Yamaguchi T, Sano K, Takakura K, Saito I, Shinohara Y, Asano T, Yasuhara H (1998) Ebselen in acute ischemic stroke: a placebo-controlled, double-blind clinical trial. Ebselen Study Group. Stroke 29:12–17
43. Saito I, Asano T, Sano K, Takakura K, Abe H, Yoshimoto T, Kikuchi H, Ohta T, Ishibashi S (1998) Neuroprotective effect of an antioxidant, ebselen, in patients with delayed neurological deficits after aneurysmal subarachnoid hemorrhage. Neurosurgery 42:269–277
44. Ogawa A, Yoshimoto T, Kikuchi H, Sano K, Saito I, Yamaguchi T, Yasuhara H (1999) Ebselen in acute middle cerebral artery occlusion: a placebo-controlled, double-blind clinical trial. Cerebrovasc Dis 9:112–118

45. Lynch E, Kil J (2009) Development of ebselen, a glutathione peroxidase mimic, for the prevention and treatment of noise-induced hearing loss. Semin Hear 30:47–55
46. Masaki C, Sharpley AL, Godlewska BR, Berrington A, Hashimoto T, Singh N, Vasudevan SR, Emir UE, Churchill GC, Cowen PJ (2016) Effects of the potential lithium-mimetic, ebselen, on brain neurochemistry: a magnetic resonance spectroscopy study at 7 tesla. Psychopharmacology 233:1097–1104
47. Beckman JA, Goldfine AB, Leopold JA, Creager MA (2016) Ebselen does not improve oxidative stress and vascular function in patients with diabetes: a randomized, crossover trial. Am J Physiol Heart Circ Physiol 311:H1431–H1436
48. Parnham MJ, Sies H (2013) The early research and development of ebselen. Biochem Pharmacol 86:1248–1253
49. Wang L, Fu J, Wang J, Jin C, Ren X, Tan Q, Li J, Yin H, Xiong K, Wang T, Liu X, Zeng H (2011) Selenium-containing thioredoxin reductase inhibitor ethaselen sensitizes non-small cell lung cancer to radiotherapy. Anti-Cancer Drugs 22:732–740
50. Ye SF, Yang Y, Wu L, Ma WW, Zeng HH (2017) Ethaselen: a novel organoselenium anticancer agent targeting thioredoxin reductase 1 reverses cisplatin resistance in drug-resistant K562 cells by inducing apoptosis. J Zhejiang Univ Sci B 18:373–382
51. Zhao F, Yan J, Deng S, Lan L, He F, Kuang B, Zeng H (2006) A thioredoxin reductase inhibitor induces growth inhibition and apoptosis in five cultured human carcinoma cell lines. Cancer Lett 236:46–53
52. Xing F, Li S, Ge X, Wang C, Zeng H, Li D, Dong L (2008) The inhibitory effect of a novel organoselenium compound BBSKE on the tongue cancer Tca8113 in vitro and in vivo. Oral Oncol 44:963–969
53. Moutet M, D'Alessio P, Malette P, Devaux V, Chaudière J (1998) Glutathione peroxidase mimics prevent TNFα- and neutrophil-induced endothelial alterations. Free Radic Biol Med 25:270–281
54. Asaf R, Blum S, Miller-Lotan R, Levy A (2007) BXT-51072 and the prevention of myocardial ischemia-reperfusion injury. Lett Drug Des Discov 4:160–162
55. Blum S, Asaf R, Guetta J, Miller-Lotan R, Asleh R, Kremer R, Levy NS, Berger FG, Aronson D, Fu X, Zhang R, Hazen SL, Levy AP (2007) Haptoglobin genotype determines myocardial infarct size in diabetic mice. J Am Coll Cardiol 49:82–87
56. Santoro S, Azeredo JB, Nascimento V, Sancineto L, Braga AL, Santi C (2014) The green side of the moon: ecofriendly aspects of organoselenium chemistry. RSC Adv 4:31521–31535
57. Bartolini D, Sancineto L, Fabro de Bem A, Tew KD, Santi C, Radi R, Toquato P, Galli F (2017) Chapter ten: Selenocompounds in cancer therapy: an overview. Adv Cancer Res 136:259–302
58. Noguchi N (2016) Ebselen, a useful tool for understanding cellular redox biology and a promising drug candidate for use in human diseases. Arch Biochem Biophys 595:109–112
59. Battin EE, Brumaghim J (2009) Antioxidant activity of sulfur and selenium: a review of reactive oxygen species scavenging, glutathione peroxidase, and metal-binding antioxidant mechanisms. Cell Biochem Biophys 55:1–23
60. Kraus RJ, Foster SJ, Ganther HE (1983) Identification of selenocysteine in glutathione peroxidase by mass spectroscopy. Biochemistry 22:5853–5858
61. Flohé L, Toppo S, Cozza G, Ursini F (2011) A comparison of thiol peroxidase mechanisms. Antioxid Redox Signal 15:763–780
62. Bhabak KP, Mugesh G (2010) Functional mimics of glutathione peroxidase: bioinspired synthetic antioxidants. Acc Chem Res 43:1408–1419
63. Caldwell KA, Tappel AL (1964) Reactions of seleno- and sulfoamino acids with hydroperoxides. Biochemistry 3:1643–1647
64. Caldwell KA, Tappel AL (1965) Acceleration of sulfhydryl oxidations by selenocystine. Arch Biochem Biophys 112:196–200
65. Paglia DE, Valentine WN (1967) Studies on the quantitative and qualitative characterization of erythrocyte glutathione peroxidase. J Lab Clin Investig 70:158–169

66. Yasuda K, Watanabe H, Yamazaki S, Toda S (1980) Glutathione peroxidase activity of D,L-selenocystine and selenocystamine. Biochem Biophys Res Commun 96:243–249
67. Wilson SR, Zucker PA, Huang RRC, Spector A (1989) Development of synthetic compounds with glutathione peroxidase activity. J Am Chem Soc 111:5936–5939
68. Engman L, Stern D, Cotgreave IA, Andersson CM (1992) Thiol peroxidase-activity of diaryl ditellurides as determined by a H1 NMR method. J Am Chem Soc 114:9737–9743
69. Bell IM, Hilvert D (1993) Peroxide dependence of the semisynthetic enzyme selenosubtilisin. Biochemistry 32:13969–13973
70. Iwaoka M, Tomoda S (1994) A model study on the effect of an amino group on the antioxidant activity of glutathione peroxidase. J Am Chem Soc 116:2557–2561
71. Hildebraunt AG, Roots I (1975) Reduced nicotinamide adenine dinucleotide phosphate (NADPH)-dependent formation and breakdown of hydrogen peroxide during mixed function oxidation reactions in liver microsomes. Arch Biochem Biophys 171:385–397
72. Haenen GR, De Rooij BM, Vermeulen NP, Bast A (1990) Mechanism of the reaction of ebselen with endogenous thiols: dihydrolipoate is a better cofactor than glutathione in the peroxidase activity of ebselen. Mol Pharmacol 37:412–422
73. Iwaoka M, Kumakura F (2008) Applications of water-soluble selenides and selenoxides to protein chemistry. Phosphorus Sulfur Silicon Relat Elem 183:1009–1017
74. Kumakura F, Mishra B, Priyadarsini KI, Iwaoka M (2010) A water-soluble cyclic selenide with enhanced glutathione peroxidase-like catalytic activities. Eur J Org Chem 2010(3):440–445
75. Tidei C, Piroddi M, Galli F, Santi C (2012) Oxidation of thiols promoted by PhSeZnCl. Tetrahedron Lett 53:232–234
76. McNeil NMR, Press DJ, Mayder DM, Garnica P, Doyle LM, Back TG (2016) Enhanced glutathione peroxidase activity of water-soluble and polyethylene glycol-supported selenides, related spirodioxyselenuranes, and pincer selenuranes. J Org Chem 81:7884–7897
77. Back TG, Dyck BP (1997) A novel camphor-derived selenenamide that acts as a glutathione peroxidase mimetic. J Am Chem Soc 119:2079–2083
78. Elsherbini M, Hamama WS, Zoorob HH, Bhowmick D, Mugesh G, Wirth T (2014) Synthesis and antioxidant activities of novel chiral ebselen analogues. Heteroat Chem 25:320–325
79. Wendel A, Fausel M, Safayhi H, Tiegs G, Otter R (1984) A novel biologically active seleno-organic compound--II. Activity of PZ 51 in relation to glutathione peroxidase. Biochem Pharmacol 33:3241–3245
80. Müller A, Gabriel H, Sies H (1985) A novel biologically active selenoorganic compound—IV. Protective glutathione-dependent effect of PZ 51 (ebselen) against ADP-Fe induced lipid peroxidation in isolated hepatocytes. Biochem Pharmacol 34:1185–1189
81. Cotgreave IA, Sandy MS, Berggren M, Moldéus PW, Smith MT (1987) Acetylcysteine and glutathione-dependent protective effect of PZ51 (ebselen) against diquat-induced cytotoxicity in isolated hepatocytes. Biochem Pharmacol 36:2899–2904
82. Back TG, Moussa Z (2002) Remarkable activity of a novel cyclic seleninate ester as a glutathione peroxidase mimetic and its facile in situ generation from allyl 3-hydroxypropyl selenide. J Am Chem Soc 124:12104–12105
83. Back TG, Moussa Z (2003) Diselenides and allyl selenides as glutathione peroxidase mimetics. Remarkable activity of cyclic seleninates produced in situ by the oxidation of allyl ω-hydroxyalkyl selenides. J Am Chem Soc 125:13455–13460
84. Back TG, Moussa Z, Parvez M (2004) The exceptional glutathione peroxidase-like activity of di(3-hydroxypropyl) selenide and the unexpected role of a novel spirodioxaselenanonane intermediate in the catalytic cycle. Angew Chem Int Ed 43:1268–1270
85. Sarma BK, Mugesh G (2005) Glutathione peroxidase (GPx)-like antioxidant activity of the organoselenium drug ebselen: unexpected complications with thiol exchange reactions. J Am Chem Soc 127:11477–11485
86. Bhabak KP, Mugesh G (2007) Synthesis, characterization, and antioxidant activity of some ebselen analogues. Chem Eur J 13:4594–4601

87. Shi H, Liu S, Miyake M, Liu KJ (2006) Ebselen induced C6 glioma cell death in oxygen and glucose deprivation. Chem Res Toxicol 19:655–660
88. Bhowmick D, Srivastava S, D'Silva P, Mugesh G (2015) Highly efficient glutathione peroxidase and peroxiredoxin mimetics protect mammalian cells against oxidative damage. Angew Chem Int Ed 54:8449–8453
89. Sies H, Arteel GE (2000) Interaction of peroxynitrite with selenoproteins and glutathione peroxidase mimics. Free Radic Biol Med 28:1451–1455
90. Azad GK, Tomar RS (2014) Ebselen, a promising antioxidant drug: mechanisms of action and targets of biological pathways. Mol Biol Rep 41:4865–4879
91. Nascimento V, Ferreira NL, Canto RFS, Schott KL, Waczuk EP, Sancineto L, Santi C, Rocha JBT, Braga AL (2014) Synthesis and biological evaluation of new nitrogen-containing diselenides. Eur J Med Chem 87:131–139
92. Pacuła AJ, Kaczor KB, Antosiewicz J, Janecka A, Długosz A, Janecki T, Wojtczak A, Ścianowski J (2017) New chiral ebselen analogues with antioxidant and cytotoxic potential. Molecules 22:492
93. Pacuła AJ, Kaczor KB, Wojtowicz A, Antosiewicz J, Janecka A, Długosz A, Janecki T, Ścianowski J (2016) New glutathione peroxidase mimetics—insights into antioxidant and cytotoxic activity. Bioorg Med Chem 25:126–131
94. Xie L, Zheng W, Xin N, Xie JW, Wang T, Wang ZY (2012) Ebselen inhibits iron-induced Tau phosphorylation by attenuating DMT1 up-regulation and cellular iron uptake. Neurochem Int 61:334–340
95. Kasraee B, Nikolic DS, Salomon D, Carraux P, Fontao L, Piguet V, Omrani GR, Sorg O, Saurat JH (2012) Ebselen is a new skin depigmenting agent that inhibits melanin biosynthesis and melanosomal transfer. Exp Dermatol 21:19–24
96. Mahadevan J, Parazzoli S, Oseid E, Hertzel AV, Bernlohr DA, Vallerie SN, Liu CQ, Lopez M, Harmon JS, Robertson RP (2013) Ebselen treatment prevents islet apoptosis, maintains intranuclear Pdx-1 and MafA levels, and preserves -cell mass and function in ZDF rats. Diabetes 62:3582–3588
97. Wang X, Yun JW, Lei XG (2014) Glutathione peroxidase mimic ebselen improves glucose-stimulated insulin secretion in murine islets. Antioxid Redox Signal 20:191–203
98. Santi C, Tidei C, Scalera C, Piroddi M, Galli F (2013) Selenium containing compounds from poison to drug candidates: a review on the GPx-like activity. Curr Chem Biol 7:25–36
99. Singh VP, Singh HB, Butcher RJ (2011) Synthesis and glutathione peroxidase-like activities of isoselenazolines. Eur J Org Chem 2011:5485–5497
100. Luo Z, Sheng J, Sun Y, Lu C, Yan J, Liu A, Luo H, Huang L, Li X (2013) Synthesis and evaluation of multi-target-directed ligands against Alzheimer's disease based on the fusion of donepezil and ebselen. J Med Chem 56:9089–9099
101. Pacuła AJ, Ścianowski J, Aleksandrzak KB (2014) Highly efficient synthesis and antioxidant capacity of N-substituted benzisoselenazol-3(2H)-ones. RSC Adv 4:48959–48962
102. Balkrishna SJ, Kumar S, Azad GK, Bhakuni BS, Panini P, Ahalawat N, Tomar RS, Detty MR, Kumar S (2014) An ebselen like catalyst with enhanced GPx activity via a selenol intermediate. Org Biomol Chem 12:1215–1219
103. Wang Z, Wang Y, Li W, Liu Z, Luo Z, Sun Y, Wu R, Huang L, Li X (2015) Computer-assisted designed "selenoxy–chinolin": a new catalytic mechanism of the GPx-like cycle and inhibition of metal-free and metal-associated Aβ aggregation. Dalt Trans 44:20913–20925
104. Kumar S, Yan J, Poon J, Singh VP, Lu X, Karlsson Ott M, Engman L, Kumar S (2016) Multifunctional antioxidants: regenerable radical-trapping and hydroperoxide-decomposing ebselenols. Angew Chem Int Ed 55:3729–3733
105. Satheeshkumar K, Mugesh G (2011) Synthesis and antioxidant activity of peptide-based ebselen analogues. Chem Eur J 17:4849–4857
106. Wilson BAP, Wang H, Nacev BA, Mease RC, Liu JO, Pomper MG, Isaacs WB (2011) High-throughput screen identifies novel inhibitors of cancer biomarker - methylacyl coenzyme A racemase (AMACR/P504S). Mol Cancer Ther 10:825–838

107. Engman L, Tunek A, Hallberg M, Hallberg A (1994) Catalytic effects of glutathione peroxidase mimetics on the thiol reduction of cytochrome c. Chem Biol Interact 93:129–137
108. Mugesh G, Singh HB (2000) Synthetic organoselenium compounds as antioxidants: glutathione peroxidase activity. Chem Soc Rev 29:347–357
109. Hodage AS, Parashiva Prabhu C, Phadnis PP, Wadawale A, Priyadarsini KI, Jain VK (2012) Synthesis, characterization, structures and GPx mimicking activity of pyridyl and pyrimidyl based organoselenium compounds. J Organomet Chem 720:19–25
110. Luchese C, Brandão R, Acker CI, Nogueira CW (2012) 2,2′-Dipyridyl diselenide is a better antioxidant than other disubstituted diaryl diselenides. Mol Cell Biochem 367:153–163
111. Nogueira C, Soares F, Nascimento P, Muller D, Rocha JBT (2003) 2,3-Dimercaptopropane-1-sulfonic acid and meso-2,3-dimercaptosuccinic acid increase mercury- and cadmium-induced inhibition of δ-aminolevulinate dehydratase. Toxicology 184:85–95
112. Singh VP, Poon J, Butcher RJ, Engman L (2014) Pyridoxine-derived organoselenium compounds with glutathione peroxidase-like and chain-breaking antioxidant activity. Chem Eur J 20:12563–12571
113. Singh VP, Poon J, Butcher RJ, Lu X, Mestres G, Ott MK, Engman L (2015) Effect of a bromo substituent on the glutathione peroxidase activity of a pyridoxine-like diselenide. J Org Chem 80:7385–7395
114. Parashiva Prabhu C, Phadnis PP, Wadawale AP, Indira Priyadarsini K, Jain VK (2012) Synthesis, characterization, structures and antioxidant activity of nicotinoyl based organoselenium compounds. J Organomet Chem 713:42–50
115. Prabhu P, Singh BG, Noguchi M, Phadnis PP, Jain VK, Iwaoka M, Priyadarsini KI (2014) Stable selones in glutathione-peroxidase-like catalytic cycle of selenonicotinamide derivative. Org Biomol Chem 12:2404–2412
116. Rafique J, Saba S, Canto R, Frizon T, Hassan W, Waczuk E, Jan M, Back D, Da Rocha J, Braga A (2015) Synthesis and biological evaluation of 2-picolylamide-based diselenides with non-bonded interactions. Molecules 20:10095–10109
117. Selvakumar K, Shah P, Singh HB, Butcher RJ (2011) Synthesis, structure, and glutathione peroxidase-like activity of amino acid containing ebselen analogues and diaryl diselenides. Chem Eur J 17:12741–12755
118. Bhabak KP, Mugesh G (2009) Amide-based glutathione peroxidase mimics: effect of secondary and tertiary amide substituents on antioxidant activity. Chem Asian J 4:974–983
119. Mugesh G, Panda A, Singh HB, Punekar NS, Butcher RJ (2001) Glutathione peroxidase-like antioxidant activity of diaryl diselenides: a mechanistic study. J Am Chem Soc 123:839–850
120. Bhowmick D, Mugesh G (2012) Tertiary amine-based glutathione peroxidase mimics: some insights into the role of steric and electronic effects on antioxidant activity. Tetrahedron 68:10550–10560
121. Bhabak KP, Mugesh G (2009) Synthesis and structure-activity correlation studies of secondary- and tertiary-amine-based glutathione peroxidase mimics. Chem Eur J 15:9846–9854
122. Ibrahim M, Hassan W, Anwar J, Deobald AM, Kamdem JP, Souza DO, Rocha JBT (2014) 1-(2-(2-(2-(1-Aminoethyl)phenyl)diselanyl)phenyl)ethanamine: an amino organoselenium compound with interesting antioxidant profile. Toxicol In Vitro 28:524–530
123. Ibrahim M, Muhammad N, Naeem M, Deobald AM, Kamdem JP, Rocha JBT (2015) In vitro evaluation of glutathione peroxidase (GPx)-like activity and antioxidant properties of an organoselenium compound. Toxicol In Vitro 29:947–952
124. Nogueira CW, Rocha JBT (2011) Toxicology and pharmacology of selenium: emphasis on synthetic organoselenium compounds. Arch Toxicol 85:1313–1359
125. Mishra B, Barik A, Kunwar A, Kumbhare LB, Priyadarsini KI, Jain VK (2008) Correlating the GPx activity of selenocystine derivatives with one-electron redox reactions. Phosphorus Sulfur Silicon Relat Elem 183:1018–1025
126. Alberto EE, Soares LC, Sudati JH, Borges ACA, Rocha JBT, Braga AL (2009) Efficient synthesis of modular amino acid derivatives containing selenium with pronounced GPx-like activity. Eur J Org Chem 2009:4211–4214

127. Soares LC, Alberto EE, Schwab RS, Taube PS, Nascimento V, Rodrigues OED, Braga AL (2012) Ephedrine-based diselenide: a promiscuous catalyst suitable to mimic the enzyme glutathione peroxidase (GPx) and to promote enantioselective C–C coupling reactions. Org Biomol Chem 10:6595–6599
128. Frizon TE, Rafique J, Saba S, Bechtold IH, Gallardo H, Braga AL (2015) Synthesis of functionalized organoselenium materials: selenides and diselenides containing cholesterol. Eur J Org Chem 2015:3470–3476
129. Kumar S, Johansson H, Engman L, Valgimigli L, Amorati R, Fumo MG, Pedulli GF (2007) Regenerable chain-breaking 2,3-dihydrobenzo[B]selenophene-5-ol antioxidants. J Org Chem 72:2583–2595
130. Hodage AS, Phadnis PP, Wadawale A, Priyadarsini KI, Jain VK (2011) Synthesis, characterization and structures of 2-(3,5-dimethylpyrazol-1-yl)ethylseleno derivatives and their probable glutathione peroxidase (GPx) like activity. Org Biomol Chem 9:2992–2998
131. Iwaoka M, Arai K (2013) From sulfur to selenium. A new research arena in chemical biology and biological chemistry. Curr Chem Biol 7:2–24
132. Nascimento V, Alberto EE, Tondo DW, Dambrowski DW, Detty MR, Nome F, Braga AL (2012) GPx-like activity of selenides and selenoxides: experimental evidence for the involvement of hydroxy perhydroxy selenane as the active species. J Am Chem Soc 134:138–141
133. Chakraborty S, Yadav SK, Subramanian M, Iwaoka M, Chattopadhyay S (2014) Dl-Trans-3,4-dihydroxy-1-selenolane (DHSred) heals indomethacin-mediated gastric ulcer in mice by modulating arginine metabolism. Biochim Biophys Acta Gen Subj 1840:3385–3392
134. Brigelius-Flohé R, Maiorino M (2013) Glutathione peroxidases. Biochim Biophys Acta Gen Subj 1830:3289–3303
135. Iwaoka M, Katakura A, Mishima J, Ishihara Y, Kunwar A, Priyadarsini K (2015) Mimicking the lipid peroxidation inhibitory activity of phospholipid hydroperoxide glutathione peroxidase (GPx4) by using fatty acid conjugates of a water-soluble selenolane. Molecules 20:12364–12375
136. Iwaoka M, Sano N, Lin YY, Katakura A, Noguchi M, Takahashi K, Kumakura F, Arai K, Singh BG, Kunwar A, Priyadarsini KI (2015) Fatty acid conjugates of water-soluble (±)-trans-selenolane-3,4-diol: effects of alkyl chain length on the antioxidant capacity. Chembiochem 16:1226–1234
137. Press DJ, Back TG (2016) The role of methoxy substituents in regulating the activity of selenides that serve as spirodioxyselenurane precursors and glutathione peroxidase mimetics. Can J Chem 94:305–311
138. Menichetti S, Capperucci A, Tanini D, Braga AL, Botteselle GV, Viglianisi C (2016) One-pot access to benzo[b][1,4]selenazines from 2-aminoaryl diselenides. Eur J Org Chem 2016:3097–3102
139. Prasad P, Singh H, Butcher R (2015) Cyclohexene-fused selenuranes and related derivatives. Molecules 20:12670–12685
140. Braverman S, Cherkinsky M, Kalendar Y, Jana R, Sprecher M, Goldberg I (2013) Synthesis of water-soluble vinyl selenides and their high glutathione peroxidase (GPx)-like antioxidant activity. Synthesis (Stuttg) 46:119–125
141. Lamani DS, Bhowmick D, Mugesh G (2012) Spirodiazaselenuranes: synthesis, structure and antioxidant activity. Org Biomol Chem 10:7933–7943
142. McNeil N, McDonnell C, Hambrook M, Back TG (2015) Oxidation of disulfides to thiolsulfinates with hydrogen peroxide and a cyclic seleninate ester catalyst. Molecules 20:10748–10762
143. Press DJ, McNeil NMR, Hambrook M, Back TG (2014) Effects of methoxy substituents on the glutathione peroxidase-like activity of cyclic seleninate esters. J Org Chem 79:9394–9401
144. Bayse C, Shoaf A (2015) Effect of methoxy substituents on the activation barriers of the glutathione peroxidase-like mechanism of an aromatic cyclic seleninate. Molecules 20:10244–10252

145. Feld JJ, Jacobson IM, Sulkowski MS, Poordad F, Tatsch F, Pawlotsky JM (2017) Ribavirin revisited in the era of direct-acting antiviral therapy for hepatitis C virus infection. Liver Int 37:5–18
146. Kirsi JJ, McKernan PA, Burns NJ, North JA, Murray BK, Robins RK (1984) Broad-spectrum synergistic antiviral activity of selenazofurin and ribavirin. Antimicrob Agents Chemother 26:466–475
147. Gebeyehu G, Marquez VE, Van Cott A, Cooney DA, Kelley JA, Jayaram HN, Ahluwalia GS, Dion RL, Wilson YA, Johns DG (1985) Ribavirin, tiazofurin, and selenazofurin: mononucleotides and nicotinamide adenine dinucleotide analogs. Synthesis, structure, and interactions with IMP dehydrogenase. J Med Chem 28:99–105
148. Weber G, Nakamura H, Natsumeda Y, Szekeres T, Nagai M (1992) Regulation of GTP biosynthesis. Adv Enzym Regul 32:57–69
149. Wray SK, Smith RH, Gilbert BE, Knight V (1986) Effects of selenazofurin and ribavirin and their 5′-triphosphates on replicative functions of influenza A and B viruses. Antimicrob Agents Chemother 29:67–72
150. Sidwell RW, Huffman JH, Call EW, Alaghamandan H, Dan Cook P, Robins RK (1986) Effect of selenazofurin on influenza A and B virus infections of mice. Antivir Res 6:343–353
151. Franchetti P, Cappellacci L, Sheikha GA, Jayaram HN, Gurudutt VV, Sint T, Schneider BP, Jones WD, Goldstein BM, Perra G, De Montis A, Loi AG, La Colla P, Grifantini M (1997) Synthesis, structure, and antiproliferative activity of selenophenfurin, an inosine 5′-monophosphate dehydrogenase inhibitor analogue of selenazofurin. J Med Chem 40:1731–1737
152. Morrey JD, Smee DF, Sidwell RW, Tseng C (2002) Identification of active antiviral compounds against a New York isolate of West Nile virus. Antivir Res 55:107–116
153. UNAIDS (2016) Global HIV statistics: Fact sheet November 2016
154. Goudgaon NM, Schinazi RF (1991) Activity of acyclic 6-(phenylselenenyl)pyrimidine nucleosides against human immunodeficiency viruses in primary lymphocytes. J Med Chem 34:3305–3309
155. Dworkin BM, Rosenthal WS, Wormser GP, Weiss L, Nunez M, Joline C, Herp A (1988) Abnormalities of blood selenium and glutathione peroxidase activity in patients with acquired immunodeficiency syndrome and AIDS-related complex. Biol Trace Elem Res 15:167–177
156. Goudgaon NM, McMillan PF, Schinazi RF (1992) 1-(Ethoxymethyl)-6-(phenylselenenyl) pyrimidines with activity against human immunodeficiency virus types 1 and 2. Antivir Chem Chemother 3:263–266
157. Nguyen MH, Schinazi RF, Shi C, Goudgaon NM, McKenna PM, Mellors JW (1994) Resistance of human immunodeficiency virus type 1 to acyclic 6-phenylselenenyl- and 6-phenylthiopyrimidines. Antimicrob Agents Chemother 38:2409–2414
158. Ni L, Schinazi RF, Boudinot FD (1995) Pharmacokinetics and toxicity of the human immunodeficiency virus inhibitor 1-ethoxymethyl-6-phenylselenenyl-5-ethyluracil in rodents. Antivir Res 27:39–47
159. Du J, Surzhykov S, Lin JS, Newton MG, Cheng YC, Schinazi RF, Chu CK (1997) Synthesis, anti-human immunodeficiency virus and anti-hepatitis B virus activities of novel oxaselenolane nucleosides. J Med Chem 40:2991–2993
160. Chu CK, Ma L, Olgen S, Pierra C, Du J, Gumina G, Gullen E, Cheng YC, Schinazi RF (2000) Synthesis and antiviral activity of oxaselenolane nucleosides. J Med Chem 43:3906–3912
161. Jeong LS, Choi YN, Tosh DK, Choi WJ, Kim HO, Choi J (2008) Design and synthesis of novel 2′,3′-Dideoxy-4′-selenonucleosides as potential antiviral agents. Bioorg Med Chem 16:9891–9897
162. Zhan P, Liu X, Fang Z, Pannecouque C, De Clercq E (2009) 1,2,3-Selenadiazole thioacetanilides: synthesis and anti-HIV activity evaluation. Bioorg Med Chem 17:6374–6879
163. Baba M (1997) Cellular factors as alternative targets for inhibition of HIV-1. Antivir Res 33:141–152
164. Thenin-Houssier S, de Vera IMS, Pedro-Rosa L, Brady A, Richard A, Konnick B, Opp S, Buffone C, Fuhrmann J, Kota S, Billack B, Pietka-Ottlik M, Tellinghuisen T, Choe H, Spicer

T, Scampavia L, Diaz-Griffero F, Kojetin DJ, Valente ST (2016) Ebselen, a small-molecule capsid inhibitor of HIV-1 replication. Antimicrob Agents Chemother 60:2195–2208
165. Sancineto L, Mariotti A, Bagnoli L, Marini F, Desantis J, Iraci N, Santi C, Pannecouque C, Tabarrini O (2015) Design and synthesis of DiselenoBisBenzamides (DISeBAs) as nucleocapsid protein 7 (NCp7) inhibitors with anti-HIV activity. J Med Chem 58:9601–9614
166. Sancineto L, Iraci N, Tabarrini O, Santi S (2017) NCp7: targeting a multitasking protein for next-generation anti-HIV drug development Part 1: Covalent inhibitors. Drug Discov Today 23:260. https://doi.org/10.1016/j.drudis.2017.10.017
167. Iraci N, Tabarrini O, Santi C, Sancineto L (2018) NCp7: targeting a multitask protein for next-generation anti-HIV drug development Part 2. Noncovalent inhibitors and nucleic acid binders. Drug Discov Today 23:687. https://doi.org/10.1016/j.drudis.2018.01.022
168. Brown JC, Newcomb WW (2011) Herpesvirus capsid assembly: insights from structural analysis. Curr Opin Virol 1:142–149
169. Elion GB (1982) Mechanism of action and selectivity of acyclovir. Am J Med 73:7–13
170. Sahu PK, Umme T, Yu J, Nayak A, Kim G, Noh M, Lee JY, Kim DD, Jeong LS (2015) Selenoacyclovir and selenoganciclovir: discovery of a new template for antiviral agents. J Med Chem 58:8734–8738
171. Sahu PK, Umme T, Yu J, Kim G, Qu S, Naik S, Jeong L (2017) Structure-activity relationships of acyclic selenopurine nucleosides as antiviral agents. Molecules 22:1167
172. Tosh DK, Choi WJ, Kim HO, Lee Y, Pal S, Hou X, Choi J, Choi S, Jeong LS (2008) Stereoselective synthesis and conformational study of novel 2′,3′-didehydro-2′,3′-dideoxy-4′-selenonucleosides. J Org Chem 73:4259–4262
173. Sartori G, Jardim NS, Marcondes Sari MH, Dobrachinski F, Pesarico AP, Rodrigues LC, Cargnelutti J, Flores EF, Prigol M, Nogueira CW (2016) Antiviral action of diphenyl diselenide on herpes simplex virus 2 infection in female BALB/c mice. J Cell Biochem 117:1638–1648
174. Sartori G, Jardim NS, Sari MHM, Flores EF, Prigol M, Nogueira CW (2017) Diphenyl diselenide reduces oxidative stress and toxicity caused by HSV-2 infection in mice. J Cell Biochem 118:1028–1037
175. Gower E, Estes C, Blach S, Razavi-Shearer K, Razavi H (2014) Global epidemiology and genotype distribution of the hepatitis C virus infection. J Hepatol 61:S45–S57
176. Cannalire R, Barreca ML, Manfroni G, Cecchetti V (2016) A journey around the medicinal chemistry of hepatitis C virus inhibitors targeting NS4B: from target to preclinical drug candidates. J Med Chem 59:16–41
177. Gastaminza P, Whitten-Bauer C, Chisari FV (2010) Unbiased probing of the entire hepatitis C virus life cycle identifies clinical compounds that target multiple aspects of the infection. Proc Natl Acad Sci 107:291–296
178. Chockalingam K, Simeon RL, Rice CM, Chen Z (2010) A cell protection screen reveals potent inhibitors of multiple stages of the hepatitis C virus life cycle. Proc Natl Acad Sci U S A 107:3764–3769
179. Wójtowicz H, Chojnacka M, Młochowski J, Palus J, Syper L, Hudecova D, Uher M, Piasecki E, Rybka M (2003) Functionalized alkyl and aryl diselenides as antimicrobial and antiviral agents: synthesis and properties. Farm 58:1235–1242
180. Wójtowicz H, Kloc K, Maliszewska I, Młochowski J, Pietka-Ottlik M, Piasecki E (2004) Azaanalogues of ebselen as antimicrobial and antiviral agents: synthesis and properties. Farm 59:863–868
181. Pietka-Ottlik M, Wójtowicz-Młochowska H, Kołodziejczyk K, Piasecki E, Młochowski J (2008) New organoselenium compounds active against pathogenic bacteria, fungi and viruses. Chem Pharm Bull (Tokyo) 56:1423–1427
182. Pietka-Ottlik M, Potaczek P, Piasecki E, Mlochowski J (2010) Crucial role of selenium in the virucidal activity of benzisoselenazol-3(2H)-ones and related diselenides. Molecules 15:8214–8228

183. Pietka-Ottlik M, Burda-Grabowska M, Woźna M, Waleńska J, Kaleta R, Zaczyńska E, Piasecki E, Giurg M (2017) Synthesis of new alkylated and methoxylated analogues of ebselen with antiviral and antimicrobial properties. ARKIVOC 2017:546–556
184. Giurg M, Gołąb A, Suchodolski J, Kaleta R, Krasowska A, Piasecki E, Piętka-Ottlik M (2017) Reaction of bis[(2-chlorocarbonyl)phenyl] diselenide with phenols, aminophenols, and other amines towards diphenyl diselenides with antimicrobial and antiviral properties. Molecules 22:974
185. Carocci M, Bakkali-Kassimi L (2012) The encephalomyocarditis virus. Virulence 3:351–367
186. Novella IS, Ebendick-Corpus BE, Zarate S, Miller EL (2007) Emergence of mammalian cell-adapted vesicular stomatitis virus from persistent infections of insect vector cells. J Virol 81:6664–6668
187. Bloom G, Merrett GB, Wilkinson A, Lin V, Paulin S (2017) Antimicrobial resistance and universal health coverage. BMJ Glob Health 2:e000518
188. Donlan RM, Costerton JW (2002) Biofilms: survival mechanisms of clinically relevant microorganisms. Clin Microbiol Rev 15:167–193
189. Nozawa R, Yokota T, Fujimoto T (1989) Susceptibility of methicillin-resistant Staphylococcus aureus to the selenium-containing compound 2-phenyl-1,2-benzoisoselenazol-3(2H)-one (PZ51). Antimicrob Agents Chemother 33:1388–1390
190. Barber M (1961) Methicillin-resistant staphylococci. J Clin Pathol 14:385–393
191. Mohammad H, Thangamani S, Seleem MN (2015) Antimicrobial peptides and peptidomimetics - potent therapeutic allies for staphylococcal infections. Curr Pharm Des 21:2073–2088
192. Thangamani S, Younis W, Seleem MN (2015) Repurposing ebselen for treatment of multidrug-resistant staphylococcal infections. Sci Rep 5:11596
193. Thangamani S, Younis W, Seleem MN (2015) Repurposing clinical molecule ebselen to combat drug resistant pathogens. PLoS One 10:e0133877
194. Bender KO, Garland M, Ferreyra JA, Hryckowian AJ, Child MA, Puri AW, Solow-Cordero DE, Higginbottom SK, Segal E, Banaei N, Shen A, Sonnenburg JL, Bogyo M (2015) A small-molecule antivirulence agent for treating Clostridium difficile infection. Sci Transl Med 7:306ra148
195. Favrot L, Grzegorzewicz AE, Lajiness DH, Marvin RK, Boucau J, Isailovic D, Jackson M, Ronning DR (2013) Mechanism of inhibition of Mycobacterium tuberculosis antigen 85 by ebselen. Nat Commun 4:2748
196. Chiou J, Wan S, Chan KF, So PK, He D, Chan EW, Chan T, Wong K, Tao J, Chen S (2015) Ebselen as a potent covalent inhibitor of New Delhi metallo-β-lactamase (NDM-1). Chem Commun 51:9543–9546
197. Lu J, Vlamis-Gardikas A, Kandasamy K, Zhao R, Gustafsson TN, Engstrand L, Hoffner S, Engman L, Holmgren A (2013) Inhibition of bacterial thioredoxin reductase: an antibiotic mechanism targeting bacteria lacking glutathione. FASEB J 27:1394–1403
198. Macegoniuk K, Grela E, Palus J, Rudzińska-Szostak E, Grabowiecka A, Biernat M, Berlicki Ł (2016) 1,2-Benzisoselenazol-3(2 H)-one derivatives as a new class of bacterial urease inhibitors. J Med Chem 59:8125–8133
199. Gustafsson TN, Osman H, Werngren J, Hoffner S, Engman L, Holmgren A (2016) Ebselen and analogs as inhibitors of Bacillus anthracis thioredoxin reductase and bactericidal antibacterials targeting Bacillus species, Staphylococcus aureus and Mycobacterium tuberculosis. Biochim Biophys Acta Gen Subj 1860:1265–1271
200. Sancineto L, Piccioni M, De Marco S, Pagiotti R, Nascimento V, Braga AL, Santi C, Pietrella D (2016) Diphenyl diselenide derivatives inhibit microbial biofilm formation involved in wound infection. BMC Microbiol 16:220
201. Naik HRP, Naik HSB, Naik TRR, Naika HR, Gouthamchandra K, Mahmood R, Ahamed BMK (2009) Synthesis of novel benzo[h]quinolines: wound healing, antibacterial, DNA binding and in vitro antioxidant activity. Eur J Med Chem 44:981–989

202. Deutch CE, Arballo ME, Cooks LN, Gomes JM, Williams TM, Aboul-Fadl T, Roberts JC (2006) Susceptibility of Escherichia coli to L-selenaproline and other L-proline analogues in laboratory culture media and normal human urine. Lett Appl Microbiol 43:392–398
203. Deutch CE, Spahija I, Wagner CE (2014) Susceptibility of Escherichia coli to the toxic L-proline analogue L-selenaproline is dependent on two L-cystine transport systems. J Appl Microbiol 117:1487–1499
204. Kumar S, Sharma N, Maurya IK, Bhasin AKK, Wangoo N, Brandão P, Félix V, Bhasin KK, Sharma RK (2016) Facile synthesis, structural evaluation, antimicrobial activity and synergistic effects of novel imidazo[1,2-a]pyridine based organoselenium compounds. Eur J Med Chem 123:916–924
205. Subha Rao SD, Joseph MP, Lavi R, Macaden R (2005) Infections related to vascular catheters in a pediatric intensive care unit. Indian Pediatr 42:667–672
206. Soteropoulos P, Vaz T, Santangelo R, Paderu P, Huang DY, Tamás MJ, Perlin DS (2000) Molecular characterization of the plasma membrane H(+)-ATPase, an antifungal target in Cryptococcus neoformans. Antimicrob Agents Chemother 44:2349–2355
207. Billack B, Santoro M, Lau-Cam C (2009) Growth inhibitory action of ebselen on fluconazole-resistant Candida albicans: role of the plasma membrane H + -ATPase. Microb Drug Resist 15:77–83
208. Loreto ÉS, Nunes Mario DA, Santurio JM, Alves SH, Nogueira CW, Zeni G (2011) In vitro antifungal evaluation and structure-activity relationship of diphenyl diselenide and synthetic analogues. Mycoses 54:e572–e576
209. Rosseti IB, Wagner C, Fachinetto R, Taube Junior P, Costa MS (2011) Candida albicans growth and germ tube formation can be inhibited by simple diphenyl diselenides [(PhSe)2, (MeOPhSe)2, (P-cl-PhSe)2, (F3CPhSe)2] and diphenyl ditelluride. Mycoses 54:506–513
210. Denardi LB, Mario DAN, de Loreto ÉS, Nogueira CW, Santurio JM, Alves SH (2013) Antifungal activities of diphenyl diselenide alone and in combination with fluconazole or amphotericin B against Candida glabrata. Mycopathologia 176:165–169
211. Kolifarhood G, Raeisi A, Ranjbar M, Haghdoust AA, Schapira A, Hashemi S, Masoumi-Asl H, Mozafar Saadati H, Azimi S, Khosravi N, Kondrashin A (2017) Prophylactic efficacy of primaquine for preventing Plasmodium falciparum and Plasmodium vivax parasitaemia in travelers: a meta-analysis and systematic review. Travel Med Infect Dis 17:5–18
212. Harris MT, Walker DM, Drew ME, Mitchell WG, Dao K, Schroeder CE, Flaherty DP, Weiner WS, Golden JE, Morris JC (2013) Interrogating a hexokinase-selected small-molecule library for inhibitors of Plasmodium falciparum hexokinase. Antimicrob Agents Chemother 57:3731–3737
213. Hüther AM, Zhang Y, Sauer A, Parnham MJ (1989) Antimalarial properties of ebselen. Parasitol Res 75:353–360
214. Hide G (1999) History of sleeping sickness in East Africa. Clin Microbiol Rev 12:112–125
215. Joice AC, Harris MT, Kahney EW, Dodson HC, Maselli AG, Whitehead DC, Morris JC (2013) Exploring the mode of action of ebselen in Trypanosoma brucei hexokinase inhibition. Int J Parasitol Drugs Drug Resist 3:154–160
216. Gordhan HM, Patrick SL, Swasy MI, Hackler AL, Anayee M, Golden JE, Morris JC, Whitehead DC (2017) Evaluation of substituted ebselen derivatives as potential trypanocidal agents. Bioorg Med Chem Lett 27:537–541
217. Sundar S, Chakravarty J (2013) Leishmaniasis: an update of current pharmacotherapy. Expert Opin Pharmacother 14:53–63
218. Araújo AP, Rocha OGF, Mayrink W, Machado-Coelho GLL (2008) The influence of copper, selenium and zinc on the response to the Montenegro skin test in subjects vaccinated against American cutaneous leishmaniasis. Trans R Soc Trop Med Hyg 102:64–69
219. Baquedano Y, Alcolea V, Toro MÁ, Gutiérrez KJ, Nguewa P, Font M, Moreno E, Espuelas S, Jiménez-Ruiz A, Palop JA, Plano D, Sanmartín C (2016) Novel heteroaryl selenocyanates and diselenides as potent antileishmanial agents. Antimicrob Agents Chemother 60:3802–3812

Chapter 3
Organoselenium in Nature

Abstract Selenium, among the naturally occurring elements, is nowadays considered the most relevant for the redox homeostasis of living systems. In this chapter, its role in plants, bacteria, and humans is scholarly discussed. Some plants have the possibility to accumulate this element, thus becoming a natural source for animals and humans, in which selenium is embedded in selenoproteins, as the 21st amino acid, selenocysteine (L-Sec). The main classes of selenoenzymes (glutathione peroxidase, thioredoxin reductase, and iodothyronine deiodinases) are reported here and the molecular mechanism that characterizes their physiological action is discussed.

3.1 Organoselenium in Plants

Selenium occurs naturally in the sedimentary rocks that were formed during the quaternary period [1]. The average Se concentration in soils is 0.4 mg/kg, even if it exists in areas that can be considered both extremely poorer and richer. In this second case, we refer to seleniferous soils and should be considered that high levels of selenium can emerge because of anthropic activities. From the soil, selenium can be transported into the plants using the normal sulfate transporting systems and, in the plant, it follows the same metabolic pathways of the sulfur derivatives, being assimilated by the incorporation in organic molecules or eliminated by volatilization in the atmosphere as **DMeSe** (dimethyl selenide) and **DMeDSe** (dimethyl diselenide) (Fig. 3.1) [2]. In the soil, and more generally in the environment, selenium is present in four different oxidation states: selenate (SeO_4^{2-}), selenite (SeO_3^{2-}), elemental (Se), and selenide (Se^{2-}). The first two species are the most abundant inorganic forms and are characterized by a good mobility in the soil due to their high solubility in water. As a direct consequence, all the parameters of the soil that affect the oxidation state of selenium can influence its bioavailability. As an example, SeO_4^{2-} is more stable and available in alkaline conditions, whereas SeO_3^{2-} is normally present in all the other conditions. The presence of cations (e.g., Ca^{2+}) promotes its adsorption, whereas the anions (Cl^- or sulfate) result in an inhibition of the process [3, 4].

© Springer International Publishing AG, part of Springer Nature 2018

E. J. Lenardão et al., *New Frontiers in Organoselenium Compounds*,
https://doi.org/10.1007/978-3-319-92405-2_3

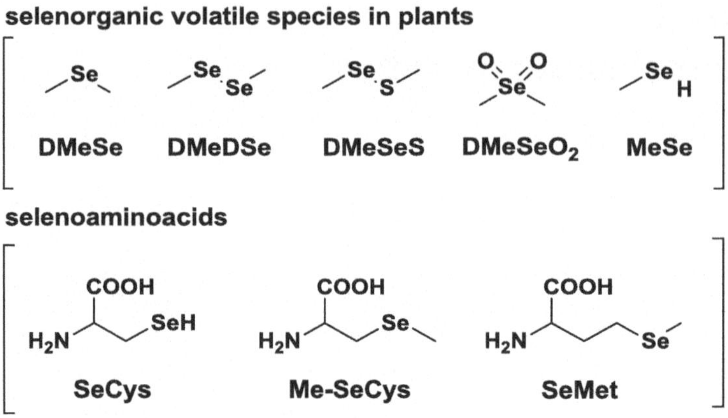

Fig. 3.1 Structures of naturally occurring organoselenium compounds

In addition, the oxidizing or reducing nature of the medium affects the distribution of selenium between the soil and the aqueous phase in a process conceptually close to the chromatography [5]. Zhao et al. reported that exist also a competition for the uptake of selenite and phosphate because they share a common transporter suggesting also a role of the uptake system of silicon in the selenite absorption [6].

Even if several studies reported the beneficial effect of selenium in plants [7–10], it is not considered an essential micronutrient as for humans. Some plants have great affinity for selenium, and for this reason they are currently named Se-accumulators [7]. Specific glutathione peroxidases (GPxs) were identified in these plants incorporating, in their active site, a cysteine in the place of a selenocysteine. These enzymes have reduced substrate specificity if compared to human GpX, thus they can act not only as glutathione peroxidase but also as thyroxine reductase [11, 12].

During the bioaccumulation, the inorganic forms of selenium are transformed into amino acids like selenomethionine (**SeMet**), selenocystine (**SeCys**), and methyl selenocystine (**MeSeCys**), or they can be methylated, leading to the formation of **DMeSe**, **DMeDSe**, dimethyl selenone (**DMeSeO$_2$**), methylselenol (**MeSeH**), and dimethyl selenyl-sulfide (**DMeSeS**) [13, 14], sometimes with the assistance of some microorganisms, such as *Alternaria* and *Penicillium corynebacterium* [15]. Selenate and selenite ions, after the uptake from the soil, are metabolized in the chloroplasts, where the first one is transformed into the second by the action of an ATP sulfurylase that affords the intermediate formation of the adenosine 5′-phosphoselenate (**APSe**), which is subsequently reduced to selenite by a specific reductase (Fig. 3.2). Even if in vitro the conversion of selenate in selenite can be easily and directly obtained by the treatment with glutathione (GSH), in vivo, the same process needs to be activated by a molecule of ATP. Once formed, selenite is reduced by GSH to selenide, which acts as co-substrate of the *O*-Acetyl-Serine (**OAcSer**) in the synthesis of the **SeCys**, that occurs still in the chloroplast. At this point, **SeCys** pass into the cytoplasm as it is, or after a series of enzymatic reactions to lead to the formation of the second main seleno-amino acid: the selenomethionine (**SeMet**) (Fig. 3.2).

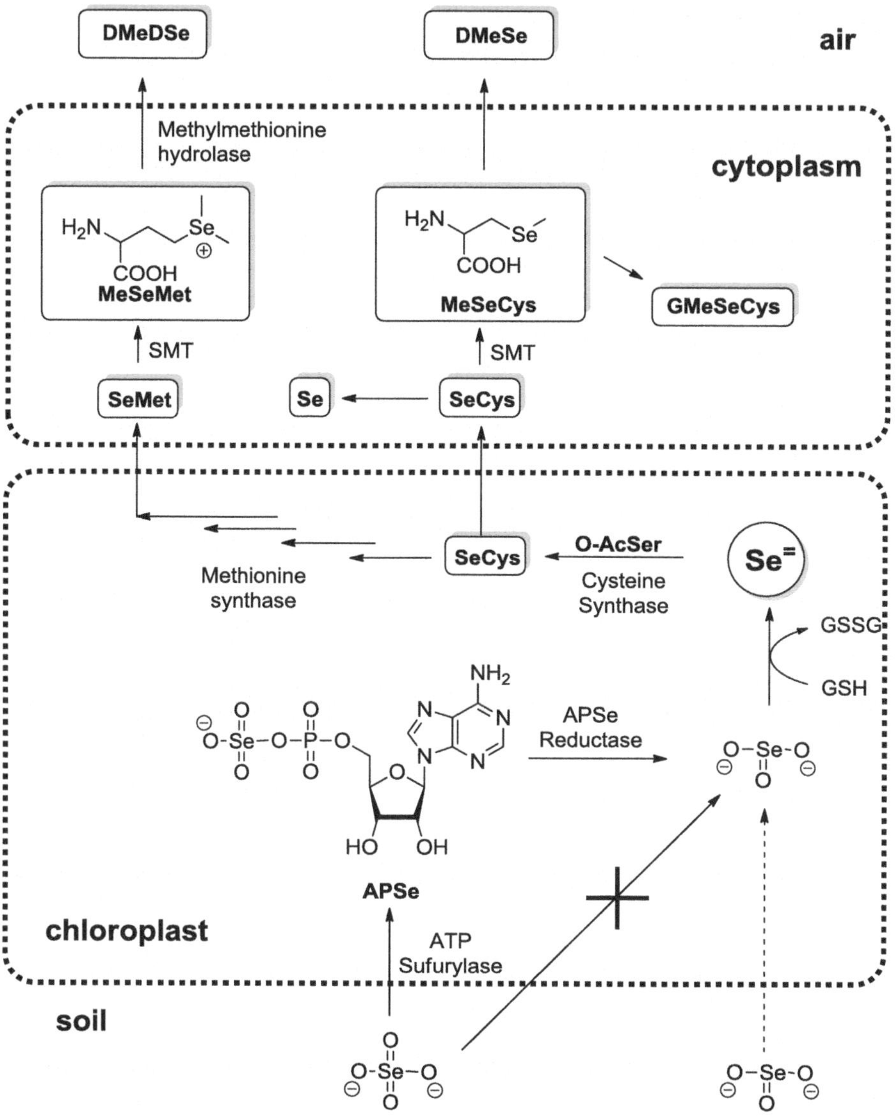

Fig. 3.2 Biosynthesis of seleno-amino acids

SeCys and **SeMet** cannot be safely stored in the plant due to the risk of their misincorporation in the proteins. In some plants, **SeCys**, by the action of a lyase, is transformed into elemental selenium. More frequently, **SeCys** and **SeMet** are methylated to obtain intermediates that, because of the impossibility to be incorporated into the proteins, can be stored in the plant, or can be hydrolyzed to the volatile form of organic selenium, that are normally released on air. In more details, **SeMet** is subjected to a methylation catalyzed by L-methionine-*S*-methyltransferase, affording methylselenomethionine (**MeSeMet**). The *S*-methyltransferase (SMT), using *S*-methylmethionine as a source of a methyl group, promotes the conversion of selenocysteine into the corresponding methylselenocysteine (**MeSeCys**). Both the methylated form **MeSeMet** and **MeSec** can be degraded to afford **DMeDSe** and **DMeSe**, respectively. In some cases, the volatilization from **MeSeMet** has been demonstrated to involve dimethyl selenopropionate as an intermediate and when it occurs, normally both mechanisms can be present at the same time. Furthermore, **MeSeCys** can be accumulated as it is or conjugated in the form of gamma-glutamyl-methylselenocysteine (**GMSeC**) [16]. Quite recently, it has been reported that some plants have the capacity to absorb organic forms of Se such as **SeCys** and **SeMet**, but not insoluble elemental Se (Se^0) or metal selenide compounds [17].

Considering the ability to accumulate selenium from the natural habitat, plants can be classified as non-Se-accumulators (<100 mg Se/kg DW), secondary-Se-accumulators (100–1000 mg Se/kg DW), and hyper-Se-accumulators (>1000 mg Se/kg DW). The latter species are normally characterized by high concentrations of selenium stored as organic **MeSec**, preferentially in young leaves and in pollen, ovules, and seeds among reproductive organs. Based on a recent theory, the ability on hyperaccumulation is a defense mechanism rapidly developed by some vegetal species that affects its interaction with herbivores, pollinators, and other plants in the neighboring area. Of course, such higher selenium content, negatively affects those partners that are selenium sensitive while facilitating the selection of the more adapted species able to survive in a seleniferous ecosystem [18].

In consideration that selenium-contaminated soils represent a potential health hazard for animals and humans (because this element rapidly enter in the food chain), the use of hyper-Se-accumulators as phytoremediators represents an eco-friendly and cost-effective strategy. The remediation occurs mainly through three mechanisms: phytoextraction, phytovolatilization, and rhizofiltration, affording Se-enriched biomass, which requires to be properly handled in terms of storage and disposal. One of the most promising uses of this biomass is the Se-biofortification of agricultural products [19]. For this purpose, it is in general necessary to select plant tissues that are edible or easily convertible into food, and that can accumulate higher and safe concentrations of Se, but not other toxic compounds [20]. Furthermore, biomass as natural selenium source could be interestingly used in the preparation of products for alimental integration in the regions with low concentration of selenium in the soil.

3.2 Selenoproteins from Bacteria to Mammals

Selenium is incorporated into selenoproteins in the form of **SeCys**, which is considered as the 21st amino acid because it is currently the unique known proteogenic *Se*-amino acid. The GPx1 was the first selenoprotein to be discovered in the rat liver, in 1978 [21]. Studies involving this enzyme have shown that the insertion of **SeCys** is codified by the codon UGA [22], which usually serves as one of the three termination codons for non-Se-protein genes.

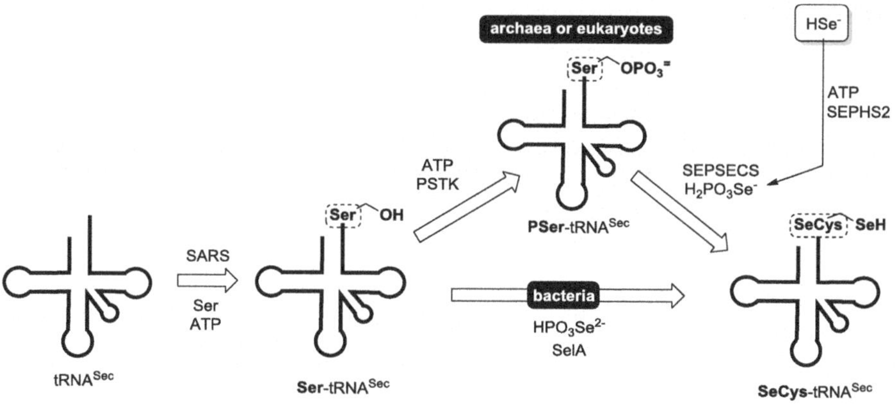

Fig. 3.3 Selenoproteins from bacteria

SeCys does not exist in cells as a free amino acid, but it is synthetized on its tRNA, with initial attachment of serine to $tRNA^{Sec}$ by seryl-tRNA synthetase (SARS), to afford the Sec-specific transfer RNA (**Ser**-$tRNA^{Sec}$). At this point, in the bacteria, **SeCys**-$tRNA^{Sec}$ is formed by the direct conversion of the OH group of serine to a selenol (SeH) group, by the action of the bacterial homodecameric enzyme selenocysteine synthase (SelA), which uses selenophosphate (HPO_3Se^{2-}) as a selenium donor [23]. In Archaea and Eukaryota, the serine residue is phosphorylated by a phosphoseryl-tRNA kinase (PSTK). Subsequently, the resulting phosphoserine (PSer), is transformed into an intermediate by Sep-tRNA:Sec-tRNA synthase (SEPSECS), and selenylated by selenophosphate to generate SeCys-$tRNA^{Sec}$ [24]. Selenophosphate derives from the reaction of selenide and ATP, catalyzed by selenophosphate synthetase 2 (SEPHS2) [25]. A multiprotein complex containing SeCys-$tRNA^{Sec}$ is bound to the selenocysteine-insertion sequence (SECIS) stem-loop in the mammalian selenoprotein mRNAs, decoding UGA SeCys codons at the ribosomal acceptor and mediating the incorporation of **SeCys** into the growing polypeptide in a process subjected to a multifactorial control (Fig. 3.3) [26].

In the human genome, 25 genes for selenoproteins have been identified even if new computational analysis were recently developed to search the SECIS sequence overcoming the complication due to the dual meaning of the UGA codon as stop and selenocysteine [27]. All the selenoproteins have a function closely correlated to the presence of the selenium atom and are generally involved in redox reactions having biological functions in redox processes, redox signaling, antioxidant defense, thyroid functionality, immune response, and their malfunctions are correlated to a series of human and animal diseases. Among all the known selenoproteins, three main classes were studied in terms of reaction mechanisms in different physiologically relevant redox processes: GPxs, thioredoxin reductases (TRxRs), and iodothyronine deiodinases (DIOs). As stated, the lack of correct functionality of these enzymes is correlated to several human diseases, such as cancer, Keshan disease, virus infections, male infertility, and abnormalities in immune responses and thyroid hormone function [28].

Table 3.1 Classification of selenoenzymes

Name	Description	Ref
GPx1	Ubiquitous cytosolic Gpx	[29]
GPx2	Gastrointestinal Gpx	[29]
GPx3	Plasma Gpx	[29]
GPx4	Ubiquitous phospholipid hydroperoxide Gpx	[29]
GPx6	Olfactory epithelium- and embryonic tissue-specific Gpx	[29]

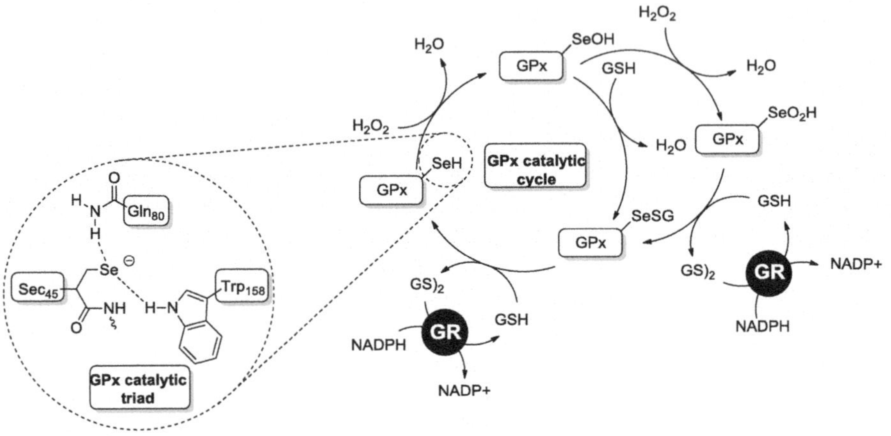

Fig. 3.4 Reaction mechanism of GPx

3.2.1 Glutathione Peroxidases (GPxs)

The family of GPx is the most important component of the antioxidant defense in mammalians. Among the eight known forms, five are demonstrated to be selenoenzymes in which the selenium of a **SeCys** is the catalytic center in the reduction of reactive species of oxygen (ROS). They are mainly classified based on the location as summarized in Table 3.1.

GPx1, GPx2, and GPx3 are homotetrameric proteins with a subunit molecular mass of 22–25 kDa and catalyze the reduction of peroxides (hydrogen peroxide and organic hydroperoxides). GPx4 is a 20–22 kDa monomeric enzyme specific for the reduction of phospholipid and cholesterol hydroperoxides, with an importance in the sperm maturation and, consequently, a role in the male fertility [30].

In the catalytic cycle of GPx (Fig. 3.4), one molecule of peroxide is reduced to water (or alcohol) consuming two molecules of glutathione (GSH), which is oxidized into the corresponding disulfide [$(GS)_2$]. The first intermediate is a selenenic acid that can be rapidly reduced by GSH affording a selenenyl sulfide, which reacts with a second molecule of cofactor GSH, regenerating the catalytic selenolate. The reducing ambient is maintained thanks to the action of the glutathione reductase and using NADPH as a cofactor. Selenium, compared to sulfur, has two main advantages: (1) being stabilized by the catalytic triad, it exists as selenol that, at physiological pH, is deprotonated; (2) it is more resistant to overoxidation. Even when it occurs, it is still possible to recover the catalytic cycle by the reduction of the possibly formed seleninic acid with glutathione [31]. In the case of sulfur, when it is subjected to overoxidation to sulfinic or sulfonic derivatives, they cannot be easily reduced back to thiols. Indeed, while sulfonic acid formation is irreversible [32], sulfinic acid was demonstrated to be reduced only in few cases by sulfiredoxin [33]. Several attempts to reproduce a GPx-like activity have been reported over the last ten years contributing to a deeper elucidation of the reaction mechanism reported in Fig. 3.4. These studies are not reported in this chapter because they are detailed in Chap. 2, besides being recently reported in several review articles [34] and book chapters [35].

3.2.2 Thioredoxin Reductases (TrxRs)

TrxRs are classified in the family of pyridine nucleotide-disulfide oxidoreductase. Nowadays, three different enzymes of this class are identified in mammals: TrxR1 in the cytosol/nucleus [36, 37], TrxR2 in mitochondria [38, 39], and TrxR3 in testis, having also glutathione and glutaredoxin reductase activity (Table 3.2) [40].

The TrxR contains a FAD-binding domain and a NADPH-binding domain and is constituted by two subunits: the *N*-terminal subunit contains a redox-active dithiol and the *C*-terminal subunit a selenothiol, representing the redox active center of the enzyme. The mechanism proposed for the catalytic activity of TrxR starts with the reduction of the *Se-S* bond on selenenylsulfide subunit, affording a selenolate that, at physiological conditions, due to the pKa of the selenol, exists as a selenium-centered anion. The reduction occurs with the consumption of a NADPH and involves the intermediate action of a molecule of FAD. At this stage, a second electron-transfer from a molecule of NADPH reduces also the disulfide subunit, generating a thiol and a free cysteine, which is stabilized by the interaction with FAD. The anionic selenium reduces the disulfide of a molecule of Trx and the reduction is completed by the attack of the neighboring thiolate. Finally, the catalytic center is regenerated by the oxidation and the formation of a disulfide in the second subunit (Fig. 3.5) [41].

TrxRs are involved in the control of cellular proliferation, viability, and apoptosis through the control of the Trx activity and redox state. TrxR is the only enzyme able to reduce oxidized Trx, providing electrons to ribonucleotide reductase, which is essential for DNA synthesis [41].

Table 3.2 Thioredoxin reductases

Name	Description	Ref
TRxR1	Cytosol/nucleus Trx	[36, 37]
TRxR2	Mitochondrial Trx	[38, 39]
TRxR3	Testis Trx	[40]

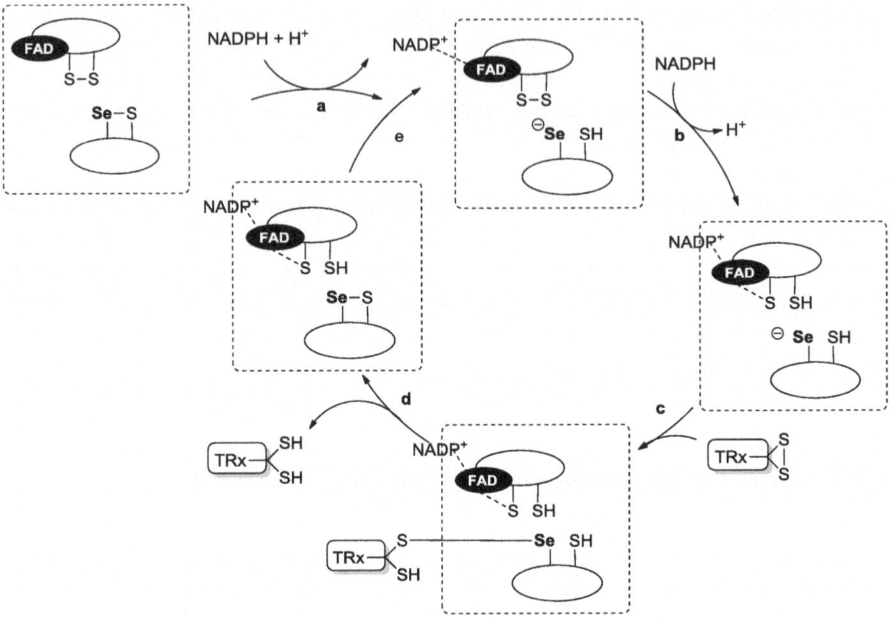

Fig. 3.5 Proposed mechanism for TrxR

Table 3.3 Iodothyronine deiodinases

Name	Description	Ref
ID-I	Inner and outer ring deiodination	[42, 43]
ID-II	Outer ring deiodination	[43]
ID-III	Inner ring deiodination	[43]

3.2.3 *Iodothyronine Deiodinases (IDs)*

The selenoenzymes classified as deiodinase are essential to control thyroid activity by the activation and deactivation of thyroid hormones. Three main classes of ID's are currently known and, besides their presence in different tissues, they have a selective interaction with the hormone, promoting a selective and reductive deiodination (Table 3.3). ID-I and ID-II are mainly involved in the activation of thyroxine (T4) into triiodothyronine (T3), increasing the thyroid activity by 5′-deiodination in the outer ring of the T4 molecule. ID-III reduces the thyroid activity by the conversion of T4 into reverse T3 (iT3) and it is also responsible for the deiodination that transforms iT3 into T2 (Fig. 3.6) [44].

The understanding of the selective deiodination mechanism is still a matter of debate and for this reason some research groups, during the last decades, proposed small-sized selenium containing derivatives as mimetics of the three isoforms of deiodinase. Mugesh and coworkers investigated a series of naphthyl-derivatives

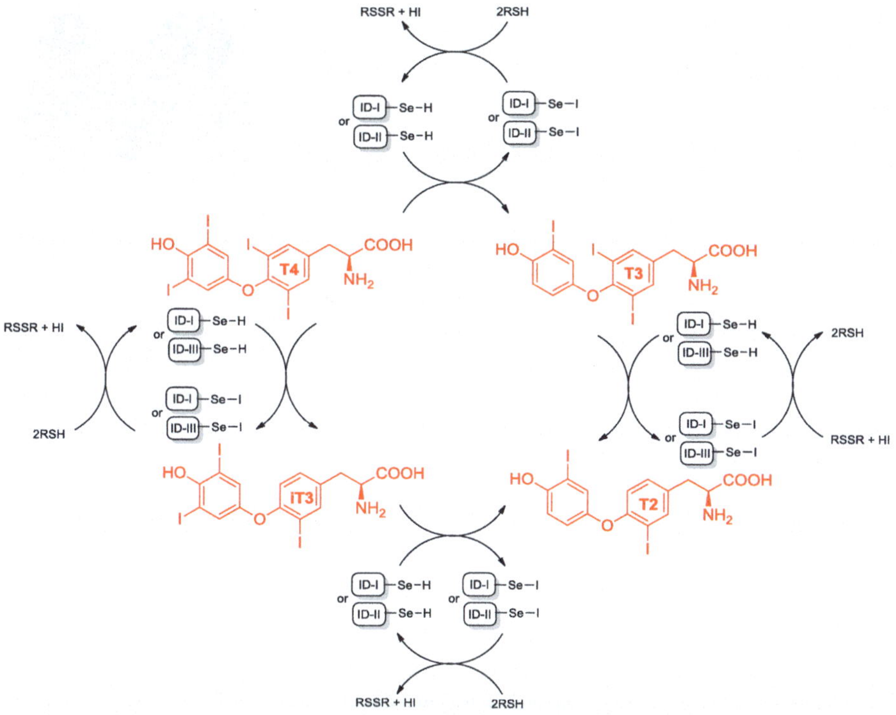

Fig. 3.6 Selective deiodination by ID-I, ID-II and ID-III

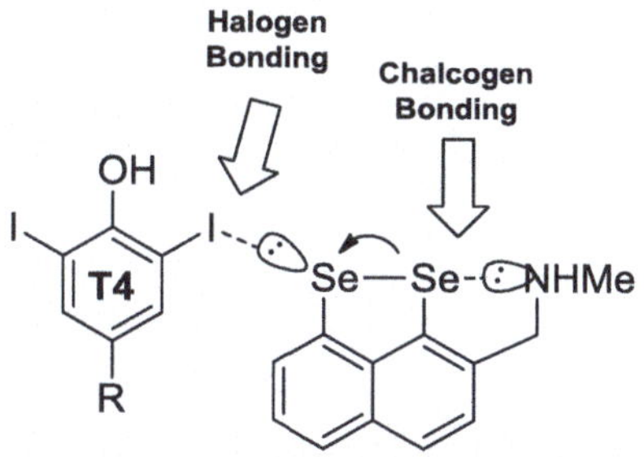

Fig. 3.7 Diselenide with deiodinase mimetic properties

functionalized as dithiol, thiol-selenol or diselenol, demonstrating the superiority of the latter based on the simultaneous presence of an intermolecular halogen bonding and an intramolecular selenium bonding (Fig. 3.7) [45–48]. The regioselectivity of these derivatives is in favor of the inner ring deioidination and, consequently, directed to reduce the thyroid function by the transformation of T4 into iT3 (mimicking the isoform III of the deiodinase).

Recently, a steric-based approach was attempted through the synthesis of hindered selenols in which the hydrophobic bulky cavity stabilizes the selenol group,

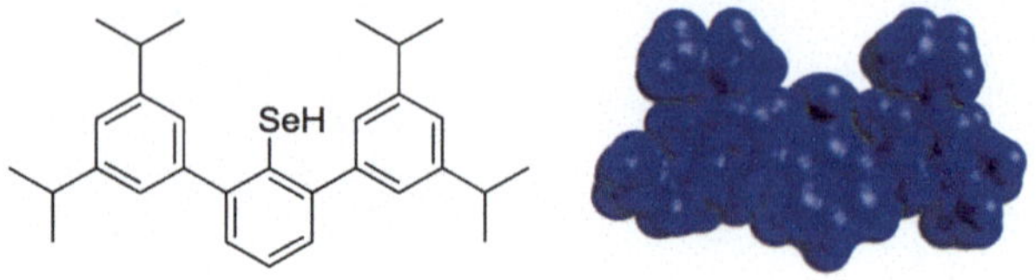

Fig. 3.8 Structure of 3,3″,5,5″-tetraisopropyl-[1,1′:3′,1″-terphenyl]-2′-selenol

as depicted in Fig. 3.8, even if it is reasonable to consider this condition still far from a real mimetic reproduction of the enzymatic cavity [49, 50].

As observed for the diselenides of Mugesh and coworkers (Fig. 3.7), in this case also an inner ring selective deiodination was observed. In consideration of the potential use as therapeutic agents in the treatment of the hypothyroidism, the synthesis of new molecules having the ability to promote the outer ring deiodination and the understanding of the different mechanisms involved in the two different deiodinations are currently highly attractive targets.

References

1. White PJ, Bowen HC, Parmaguru P, Fritz M, Spracklen WP, Spiby RE, Meacham MC, Mead A, Harriman M, Trueman LJ, Smith BM, Thomas B, Broadley MR (2004) Interactions between selenium and sulphur nutrition in Arabidopsis thaliana. J Exp Bot 55:1927–1937
2. Martens DA, Suarez DL (1997) Selenium speciation of soil/sediment determined with sequential extractions and hydride generation atomic absorption spectrophotometry. Environ Sci Technol 31:133–139
3. Hyun S, Burns PE, Murarka I, Lee LS (2006) Selenium(IV) and (VI) sorption by soils surrounding fly ash management facilities. Vadose Zone J 5:1110–1118
4. Grieve CM, Poss JA, Suarez DL, Dierig DA (2001) Lesquerella growth and selenium uptake affected by saline irrigation water composition. Ind Crop Prod 13:57–65
5. Brown KM, Arthur JR (2001) Selenium, selenoproteins and human health: a review. Public Health Nutr 4:593–599
6. Zhao XQ, Mitani N, Yamaji N, Shen RF, Ma JF (2010) Involvement of silicon influx transporter OsNIP2;1 in selenite uptake in rice. Plant Physiol 153:1871–1877
7. Shanker AK (2006) Countering UV-B stress in plants: does selenium have a role? Plant Soil 282:21–26
8. Cartes P, Jara AA, Pinilla L, Rosas A, Mora ML (2010) Selenium improves the antioxidant ability against aluminium-induced oxidative stress in ryegrass roots. Ann Appl Biol 156:297–307
9. Hasanuzzaman M, Hossain MA, Fujita M (2011) Selenium-induced up-regulation of the antioxidant defense and methylglyoxal detoxification system reduces salinity-induced damage in rapeseed seedlings. Biol Trace Elem Res 143:1704–1721
10. Saidi I, Chtourou Y, Djebali WJ (2014) Selenium alleviates cadmium toxicity by preventing oxidative stress in sunflower (Helianthus annuus) seedlings. Plant Physiol 171:85–91
11. Eshdat Y, Holland D, Faltin Z, Ben-Hayyim G (1997) Plant glutathione peroxidases. Physiol Plant 100:234–249
12. Faltin Z, Camoin L, Ben-Hayyim G, Perl A, Beeor-Tzahar T, Strosberg AD, Holland D, Eshdat Y (1998) Cysteine is the presumed catalytic residue of Citrus sinensis phospholipid hydroperoxide glutathione peroxidase over-expresses under salt stress. Physiol Plant 104:741–746
13. Hansen D, Duda PJ, Zayed A, Terry N (1998) Selenium removal by constructed wetland: role of biological volatilization. Environ Sci Technol 32:591–597

14. Frankenberger WT Jr, Karlson U (1994) Microbial volatilization of selenium from soils and sediments. In: Frankenberger WT Jr, Benson S (eds) Selenium in the environment. Marcel Dekker, New York, pp 369–387
15. Azaizeh HA, Gowthaman S, Terry N (1997) Microbial selenium volatilization in rhizosphere and bulk soils from a constructed wetland. J Environ Qual 26:666–672
16. Ellis DR, Salt DE (2003) Plants, selenium and human health. Curr Opin Plant Biol 6:273–279
17. White PJ, Broadley MR (2009) Biofortification of crops with seven mineral elements often lacking in human diets-iron, zinc, copper, calcium, magnesium, Se and iodine. New Phytol 182:49–84
18. Schiavon M, Pilon-Smits EAH (2017) The fascinating facets of plant selenium accumulation - biochemistry, physiology, evolution and ecology. New Phytol 213:1582–1596
19. Lin ZQ, Haddad S, Hong J, Morrissy J, Bañuelos GS, Zhang LY (2014) Use of selenium-contaminated plants from phytoremediation for production of selenium-enriched edible mushrooms. In: Bañuelos GS, Lin ZQ, Yin XB (eds) Selenium in the environment and human health. CRC Press, Boca Raton, pp 124–126
20. Rodrigo S, Santamaria O, Chen Y, McGrath SP, Poblaciones MJ (2014) Selenium speciation in malt, wort, and beer made from selenium biofortified two-rowed barley grain. J Agric Food Chem 62:5948–5953
21. Forstrom JW, Zakowski JJ, Tappel AL (1978) Identification of the catalytic site of rat liver glutathione peroxidase as selenocysteine. Biochemistry 17:2639–2644
22. Chambers I, Frampton J, Goldfarb P, Affara N, Mcbain W, Harrison PR (1986) The structure of the mouse glutathione peroxidase gene: the selenocysteine in the active site is encoded by the 'termination' codon, TGA. EMBO J 5:1221–1227
23. Leinfelde W, Zehelein E, Mandrand-Berthelot MA, Boeck A (1988) Gene for a novel tRNA species that accepts L-serine and cotranslationally inserts selenocysteine. Nature 331:723–725
24. Lee BJ, Worland PJ, Davis JN, Stadtman TC, Hatfield DL (1989) Identification of a selenocysteyl-tRNA (Ser) in mammalian cells that recognizes the nonsense codon UGA. J Biol Chem 264:9724–9727
25. Labunskyy VM, Hatfield DL, Gladyshev VN (2014) Selenoproteins: molecular pathways and physiological roles. Physiol Rev 9:739–777
26. Low SC, Grundner-Culemann E, Harney JW, Berry MJ (2000) SECISSBP2 interactions dictate selenocysteine incorporation efficiency and selenoprotein hierarchy. EMBO J 19:6882–6890
27. Mariotti M (2018) SECISearch3 and Seblastian: in-Silico tools to predict SECIS elements and selenoproteins. In: Chavatte L (ed) Selenoproteins, Methods in molecular biology, vol 1661. Humana Press, New York
28. Lu J, Holmgren A (2009) Selenoproteins. J Biol Chem 284:723–727
29. Kryukov GV, Castellano S, Novoselov SV, Lobanov AV, Zehtab O, Guigo R, Gladyshev VN (2003) Characterization of mammalian selenoproteomes. Science 300:1439–1443
30. Ursini F, Heim S, Kiess M, Maiorino M, Roveri A, Wissing J, Flohe L (1999) Dual function of the selenoprotein PHGPx during sperm maturation. Science 285:1393–1396
31. Snider GW, Ruggles E, Khan N, Hondal RJ (2013) Selenocysteine confers resistance to inactivation by oxidation in thioredoxin reductase: comparison of selenium and sulfur enzymes. Biochemistry 52:5472–5481
32. Yang KS, Kang SW, Woo HA, Hwang SC, Chae HZ, Kim K, Rhee SG (2002) Inactivation of human peroxiredoxin I during catalysis as the result of the oxidation of the catalytic site cysteine to cysteine-sulfinic acid. J Biol Chem 277:38029–38036
33. Woo HA, Chae HZ, Hwang SC, Yang K, Kang SW, Kim K, Rhee SG (2003) Reversing the inactivation of peroxiredoxins caused by cysteine sulfinic acid formation. Science 300:653–656
34. Barcellos A, Abenante L, Sarro M, Leo I, Lenardão EJ, Perin G, Santi C (2017) New prospective for redox modulation mediated by organoselenium and organotellurium compounds. Curr Org Chem 21:2044–2061
35. Santi C, Marini F, Lenardão EJ (2018) Looking beyond the traditional idea of glutathione peroxidase mimics as antioxidants. In: Jain VK, Priyadarsini KI (eds) Organoselenium

compounds in biology and medicine: synthesis, biological and therapeutic treatments. Royal Society of Chemistry, Cambridge, pp 35–76
36. Gladyshev VN, Jeang KT, Stadtman TC (1996) Selenocysteine, identified as the penultimate C-terminal residue in human T-cell thioredoxin reductase, corresponds to TGA in the human placental gene. Proc Natl Acad Sci U S A 93:6146–6151
37. Zhong L, Arnér ES, Ljung J, Aslund F, Holmgren A (1998) Rat and calf thioredoxin reductase are homologous to glutathione reductase with a carboxyl-terminal elongation containing a conserved catalytically active penultimate selenocysteine residue. J Biol Chem 273:8581–8591
38. Lee SR, Kim JR, Kwon KS, Yoon HW, Levine RL, Ginsburg A, Rhee SG (1999) Molecular cloning and characterization of a mitochondrial selenocysteine-containing thioredoxin reductase from rat liver. J Biol Chem 274:4722–4734
39. Biterova EI, Turanov AA, Gladyshev VN, Barycki JJ (2005) Crystal structures of oxidized and reduced mitochondrial thioredoxin reductase provide molecular details of the reaction mechanism. Proc Natl Acad Sci U S A 102:15018–15023
40. Sun QA, Kirnarsky L, Sherman S, Gladyshev VN (2001) Selenoprotein oxidoreductase with specificity for thioredoxin and glutathione systems. Proc Natl Acad Sci U S A 98:3673–3678
41. Zhong L, Arnér ESJ, Holmgren A (2000) Structure and mechanism of mammalian thioredoxin reductase: the active site is a redox-active selenothiol/selenenylsulfide formed from the conserved cysteine-selenocysteine sequence. Proc Natl Acad Sci U S A 97:5854–5859
42. Moreno M, Berry M, Horst C, Thoma R, Goglia F, Harney JW, Larsen PR, Visser TJ (1994) Activation and inactivation of thyroid hormone by type I iodothyronine deiodinase. FEBS Lett 344:143–146
43. Kaplan MM (1984) The role of thyroid hormone deiodination in the regulation of hypothalamo-pituitary function. Neuroendocrinology 38:254–260
44. Barbosa NV, Nogueira CW, Nogara PA, de Bem AF, Aschner M, Rocha JBT (2017) Organoselenium compounds as mimics of selenoproteins and thiol modifier agents. Metallomics 9:1703–1734
45. Manna D, Mugesh G (2010) A chemical model for the inner-ring deiodination of thyroxine by iodothyronine deiodinase. Angew Chem Int Ed 49:9246–9249
46. Manna D, Mugesh G (2011) Deiodination of thyroid hormones by iodothyronine deiodinase mimics: does an increase in the reactivity alter the regioselectivity? J Am Chem Soc 133:9980–9983
47. Manna D, Mugesh G (2012) Regioselective deiodination of thyroxine by iodothyronine deiodinase mimics: an unusual mechanistic pathway involving cooperative chalcogen and halogen bonding. J Am Chem Soc 134:4269–4279
48. Mondal S, Mugesh G (2014) Regioselective deiodination of iodothyronamines, endogenous thyroid hormone derivatives, by deiodinase mimics. Chemistry 20:11120–11128
49. Goto K, Sonoda D, Shimada K, Sase S, Kawashima T (2010) Modeling of the 5′-deiodination of thyroxine by iodothyronine deiodinase: chemical corroboration of a selenenyl iodide intermediate. Angew Chem Int Ed 49:545–547
50. Sase S, Kakimoto R, Kimura R, Goto K (2015) Synthesis of a stable primary-alkyl-substituted selenenyl iodide and its hydrolytic conversion to the corresponding selenenic acid. Molecules 20:21415–21420

Chapter 4
Nonbonded Interaction: The Chalcogen Bond

Abstract Compared to canonical weak, nonbonded interactions, the research on the chalcogen bond (CB) is still in its infancy even if it can be easily considered fascinating. In this chapter a brief summary of its application is given. A general introductory section is detailed to place the CB in the context of weak interactions. Then, such nonbonded interactions are considered from the proteomic perspective, followed by a section focused on the chalcogen bond in drug discovery processes. Its impact on organic synthesis is presented together with few examples of chalcogen bond-assisted catalysis. The conclusive section is devoted to the recently reported examples of CBs in material chemistry.

4.1 General Introduction: Noncovalent Interactions

Starting from their identification in late 1800 by Johannes Diderik van der Waals, noncovalent interactions are thought to play a major role in several research fields related to chemistry. They are indeed important in synthesis, in catalysis, in medicinal chemistry, and in the design of new materials and drugs [1–3]. In this chapter we refer to this particular kind of chemical bonds using both the term noncovalent and nonbonded interactions.

From a general point of view, nonbonded interactions are weaker if compared to covalent bonds, indeed their bond energies usually varies from 1 to 5 kcal/mol; at room temperature, such energy barrier is usually overcome by the kinetic energy of the molecules, which break and restore these bonds continuously. Their transient nature, at ambient temperature, accounts for their description as interactions rather than bonds. Despite their weakness, multiple nonbonded interactions play together to produce highly stable and specific associations between different molecules, especially macromolecules such as proteins or polymeric materials.

Besides the well-known ionic interactions, hydrogen bonds, van der Waals forces, π-π interactions, cation-π and anion-π interactions, other, less-conventional, noncovalent bonds have been described. Among these is the ***halogen*** bond that is established when an electrophilic halogen is interacting with a Lewis base (Fig. 4.1a) [4]. Similarly, the ***pnicogen*** bond (Fig. 4.1b) is established when the

© Springer International Publishing AG, part of Springer Nature 2018

E. J. Lenardão et al., *New Frontiers in Organoselenium Compounds*,
https://doi.org/10.1007/978-3-319-92405-2_4

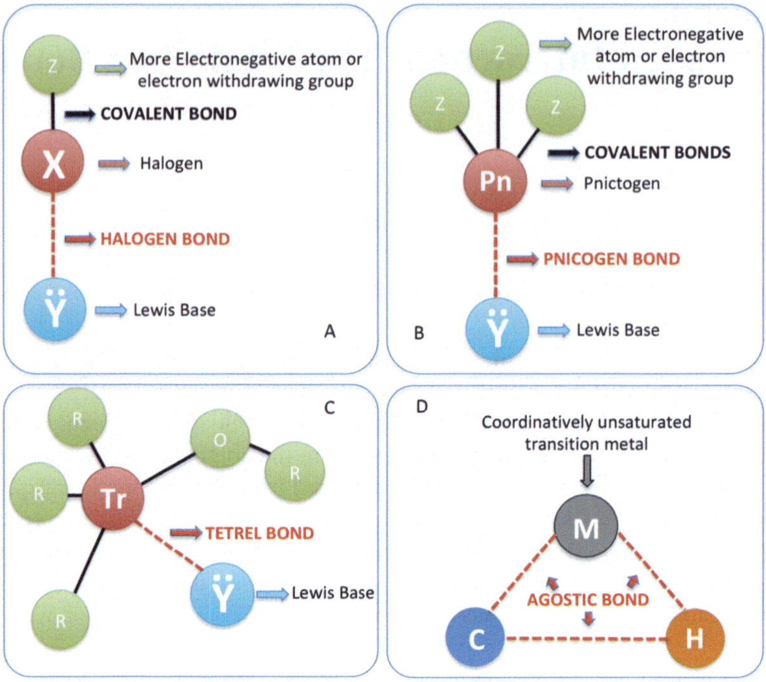

Fig. 4.1 Noncovalent interactions

Lewis acid is a member of the pnictide family, which includes phosphorus, arsenic, antimony, and bismuth [5]. When the Lewis acid is a group 13 element, then the noncovalent interaction is called ***icosagen*** bond (not shown in Fig. 4.1, but it is similar to halogen and pnicogen bonds) [6]. In 2013, the term ***tetrel*** bond (Fig. 4.1c) was suggested by Frontera and coworkers to describe the noncovalent interaction in which the electrophilic actor is one of the group 14 elements [7]. Finally, the ***aerogen*** bond was also defined, for noble gas-centered Lewis acids [8]. In the organometallic research field, the ***agostic*** bond (Fig. 4.1d) is very important. It is established in many transition states of transition metal catalyzed reaction and is formed by a coordinatively unsaturated transition metal and a C-H bond, while the ***anagostic*** version is similar, but possesses a more electrostatic character [9].

Even if the covalent bonds determine the primary structure of molecules, the abovementioned interactions surely influence their tertiary and quaternary arrangement, together with their ability to interact with macromolecular partners, their shape, and thus their functions. For these reasons, nonbonding interactions play a major role in a number of biochemical processes controlling pivotal mechanisms of living systems [10].

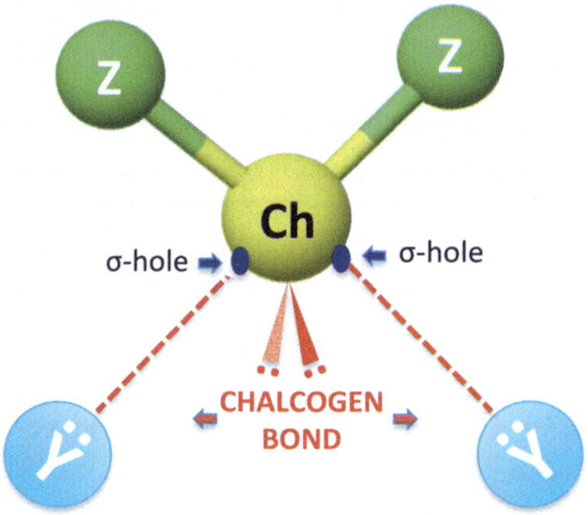

Fig. 4.2 Schematic representation of chalcogen bonds

4.2 Insights into the Chalcogen Bond

The chalcogen bonds (CBs), together with the abovementioned hydrogen, aerogen, halogen, pnicogen, tetrel, and icosagen bonds, compose the so-called σ-hole directed interactions. The term σ-hole appeared in the literature in 2007 and describes the lower electronic densities that are found on the extensions of σ bonds [11]. Usually, there is a positive electrostatic potential associated with this lower electronic density, through which attractive interactions with negative sites occur. This served to explain the Coulombic nature of the halogen bonding, which was discovered earlier [12]. The readers interested in studying in depth the σ-hole concept and its implications in several chemistry-related research field, are directed to the recently reported review of Resnati and coworkers [13].

For halogens, Politzer proved the existence of the σ-hole, which looked like an electropositive crown directed to the electron donor, at the outermost region of the atom. At the same time, the three lone electron pairs produce a belt of negative electrostatic potential around the central part of the halogen atom, leaving the possibility to act as an electron donor [14]. Essentially the same is for the chalcogens, even if they have just two lone pairs instead of three. As a general rule, the size of the σ-hole increases as the polarizability does while a reverse relationship was found with the electronegativity. Thus, for group 16 atoms, the σ-hole size increases from oxygen to the heavier tellurium, which is able to make the strongest CBs. A key feature of the CB is that it is tunable; its strength can be modified by varying the electronic nature of the Z group which is attached to the chalcogen itself (Fig. 4.2). In their bivalent form, the chalcogen σ-holes are localized along the axes of the covalent bonds and in that position the CB occurs (Fig. 4.2). At the same time, the Z–Ch covalent bonds elongate upon the formation of the CB [15]. The presence of two lone pairs makes the σ-holes flanked by a negative electrostatic potential, which explains the high directionality of the CB. The typical Z-Ch—Y angle is, indeed, close to 180°. A comparison with the parent hydrogen bond reveals that the CB is hydrophobic and more sensitive to the steric hindrance, because of the larger size which characterizes the group 16 atoms [16].

The CB attracted great attention due to its applicability in biochemistry, polymer science, crystal engineering, and supramolecular chemistry [17, 18]. Here the role of the chalcogen bond is discussed in the frame of proteomic, drug design, organic synthesis, catalysis, and material chemistry. The dissertation is not intended to be exhaustive, but it is meant to give a brief glance at the potentialities and perspectives of this relatively new, nonbonded interaction.

4.3 Chalcogen Bond in Proteins

Weak, nonbonded interactions are of pivotal importance in protein structures and functions. Ionic interactions, hydrogen bonds, and van der Waals forces play the major role, while some uncommon interaction patterns, such as C–H···*O* hydrogen bonds [19], cation-π interactions [20], and CH/π hydrogen bond [21], have recently been characterized as essential for the folded protein structures and enzyme catalysis. The chalcogen bond clearly belongs to this category, being identified in several protein structures just recently [22].

The CB most commonly observed in peptides and proteins is between sulfur and various nucleophiles, this is because tellurium is practically absent in living systems and selenium takes part only of a limited number of proteins (see Chaps. 2 and 3 for a discussion in the role of selenium in biochemical processes). Oxygen, being the most electronegative and the least polarizable, has a poorly represented σ-hole; thus, it acts as nucleophile rather than CB acceptor. Cysteine, cystine and methionine are the sulfur-containing amino acids engaged in CBs, mostly having as electron-donor partners the carbonyl groups of amidic backbones. Interestingly, the directionality of the sulfur-centered CB found in proteins is different from that observed in small molecules. In both cases, sulfur approaches the CB donor through its σ-hole, but in the case of small organic molecules, the CB lies on the plane described by the carbonyl (the CB can be described to as $n_O \rightarrow \sigma^*_S$) (Fig. 4.3a), while in proteins, in the S--O interaction the sulfur atom accesses vertically to the carbonyl plane ($\pi_O \rightarrow \sigma^*_S$) (Fig. 4.3b) [22].

The *S*--*O*=C interaction with cystine is mainly found in helices where it supports the groove stability, while the thioether group of methionine makes weaker CBs with attenuated directionality. Also, *S*--*N* interactions have been found, they are geometrically similar to that established with carbonyls, whit sulfur accessing amidic nitrogen from the vertical side. On the contrary, the *S*--*S* interactions, especially between cysteine bridges, are similar to that found in small molecules, thus stabilized by $n_S \rightarrow \sigma^*_S$ interaction [23].

The presence of chalcogen bonds was demonstrated in several proteins, the first was phospholipase A_2 (PLA_2), in which four *S*--*O* and one *S*--*N* interactions are responsible for its structural stability and, laying these nonbonded interactions in the neighborhood of the active site, for the catalysis [24]. Ribonuclease A (RNase

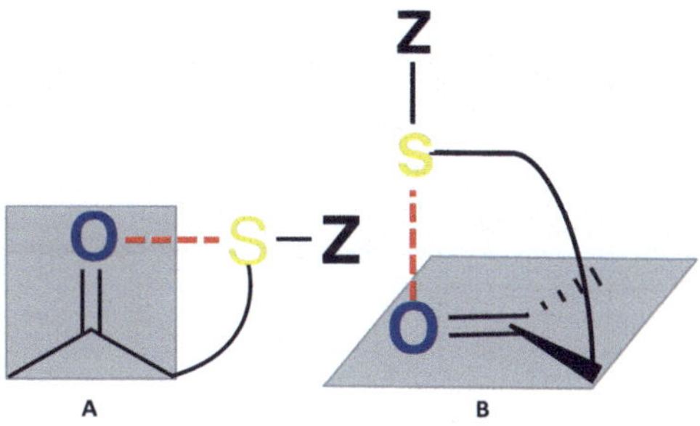

Fig. 4.3 Chalcogen bonds. (**a**) In small molecules and (**b**) in proteins

A) has two *S--O* interaction and one *S--N* interaction, while insulin features just two *S--O* interactions, one in each chain composing its structure [22]. Of course, these are just selected examples; for a deeper discussion, see Refs. [22, 25, 26].

The chalcogen bonding is not only important for the peptides structure, but also for their functions. For example, CBs were found between RNase A and its substrates [27]; the *S--O* CB of a methionine substrate and the carbonyl of Asp118 plays a major role in the enzymatic activity of *S*-adenosylmethionine synthetase [28]. In 1999, Brandt and coworkers proved that the cleavage of a disulfide bond in the extracellular region of G-protein receptors is assisted by a CB interaction between Cys121 and the carboxylic group of Asp288 [29]. *S--N* interactions between the hypothiocyanate (OSCN) ligand and the imidazole ring of His109 was found to be important for the enzymatic inhibition of lactoperoxidase [30]. Finally, an analogous CB interaction was found between the sulfenic acid form of Cys50 and the imidazole ring of His42 in the key antioxidant enzyme peroxiredoxin [31].

4.4 Chalcogen Bond in Drug Design

If compared with the canonical weak forces, such as hydrogen bonding and van der Waals interaction, the importance of chalcogen bond in drug design is underestimated and from an applicative standpoint, such an noncovalent interaction has been not fully exploited so far. A possible explanation could be that, with the exception of oxygen, among chalcogens just sulfur is present in several examples of biologically active compounds, while selenium and tellurium are poorly represented although they are better chalcogen bond acceptors with respect to the lighter homologue [32]. Fortunately, the importance of selenium in drug discovery starts to be recognized with a growing number of articles reporting the pharmacological properties of organoselenium compounds (as showed in Chap. 2) [33–38]. Sulfur is present in several clinically approved drugs; selected examples include the β-lactam-based antibiotics and their synthetic homologues, the sulfamidic class of antibacterials and the anticancer agent bleomycin, among others. In most of the cases, sulfur is present in the form of thiophene, which is conceived as the canonical bioisosteric replacer of the phenyl ring. Interestingly, in more than one example the presence of

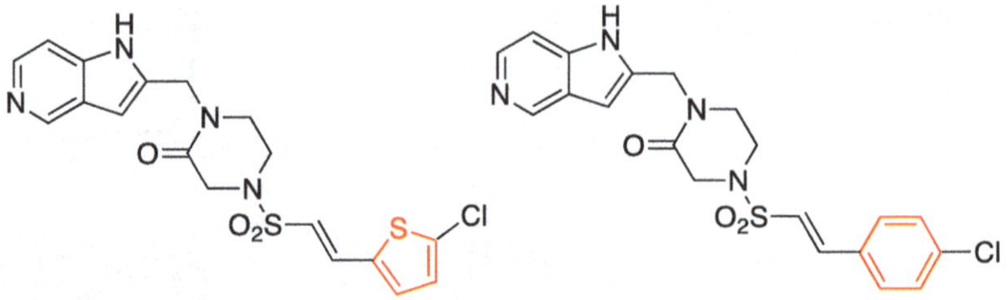

1 **2**

Fig. 4.4 Structures of Factor Xa inhibitors

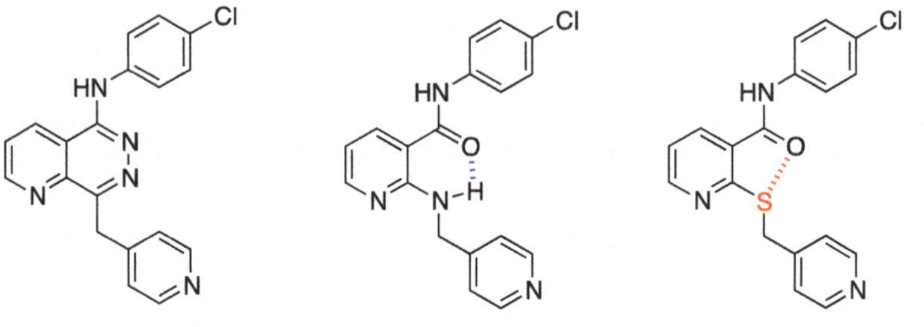

3
VEGF IC_{50} = 37 nM

4
VEGF IC_{50} = 20 nM

5
VEGF IC_{50} = 26 nM

Fig. 4.5 Structures of VEGF inhibitors

a thiophene ring endows the molecules with a superior activity if compared to the phenyl-containing counterpart. As examples, compound **1** of Fig. 4.4, together with its analogue **2**, were designed to inhibit Factor Xa. Even if they are supposed to have the same pharmacodynamic properties, i.e., the same binding mode to the pharmacological target, the thiophene-containing compound showed a 40-fold superior potency, plausibly as a result of a chalcogen bond with proximal backbone amides of the enzyme [39].

The possibility to control the conformation of a molecule improves its binding affinity thus enabling an optimal target recognition process enhancing, at the same time, the selectivity and mitigating the off-target interaction, which may result in toxicity. Although the canonical nonbonded interactions have been widely used to enrich the molecular population with the desired conformer, the intramolecular chalcogen bond is underappreciated [40]. Among chalcogens, the majority of examples regard sulfur interactions, which however have been noted in post facto analyses of crystallographic data, while relatively few examples were reported in the literature where this interaction has been exploited in a prospective fashion.

The set of vascular endothelial growth factor (VEGF) kinase inhibitors gives a noteworthy example of conformational control and effective bioisosteric relationships between the phthalazine ring found in compound **3** (Fig. 4.5), the intramolecularly *H*-bonded analogue **4** and the chalcogen bond containing derivative **5**.

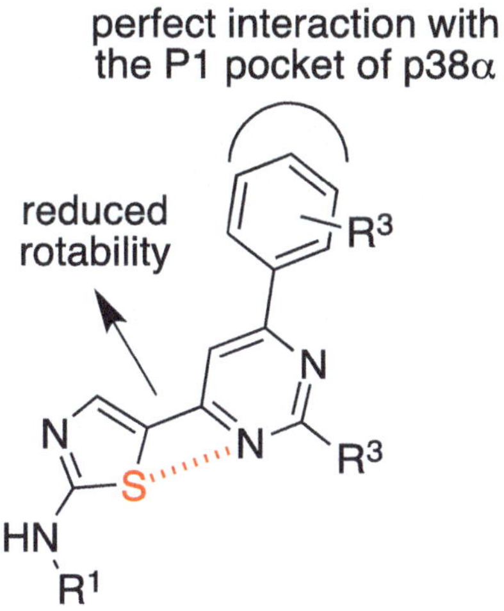

Fig. 4.6 Thiazole-based p38α MAP kinase inhibitors

Unexpectedly, the thioether **5** is as potent as compound **4**; this is because the *S* atom can be considered as an effective mimic of the *NH* group, since it is capable of stabilizing the planar conformation through the chalcogen bond with the amidic oxygen. The intramolecular non-bonded interaction was confirmed through the single-crystal X-ray analysis of compound **5**, in which the *S--O* distance is of 2.8 Å, which is lower than the sum of their van der Waals radii [41, 42].

A similar example is the development of thiazole-based p38α MAP kinase inhibitors featuring an intramolecular *S--N* chalcogen bond [43]. The inhibition of the p38 MAP kinase pathway is known to be effective in controlling the release of TNF-α and IL-1β, that are among the most prominent pro-inflammatory cytokines [44]. As a result, small molecules able to inhibit p38 attracted interest within the pharmaceutical industry due to the potential to effectively treat significant inflammatory diseases such as rheumatoid arthritis [45]. As reported in Fig. 4.6, and experimentally demonstrated by X-ray crystallographic studies, this class of p38α MAP kinase inhibitors utilizes a unique intramolecular *S--N* nonbonding interaction to reduce bond roteability and stabilize the preferred conformation for a proper binding to p38a [43].

As discussed in Chap. 2, selenazofurin (compound **6**), derived from ribavirin, showed a potent and broad antiviral activity. Together with its sulfur (thiazofurin, compound **7**) and oxygen (compound **8**) analogues, they are activated by kinases through sequential phosphorylation in order to inhibit inosine monophosphate dehydrogenase (IMPDH), a critical enzyme in the synthesis of guanosine triphosphate. For compound **7**, solid state analysis revealed a *S--O* chalcogen bond, which is present also in the alpha anomer (compound not shown) [46]. The same was found also for selenazofurin, where the *Se--O* interatomic distance was found to be of 3.0 Å, which is shorter than the 3.40 Å sum of the van der Waals radii of the two atoms [47]. Quantum mechanical calculations predicted that the electron-withdrawing properties of the C-4 carboxamide substituent increases the energy of

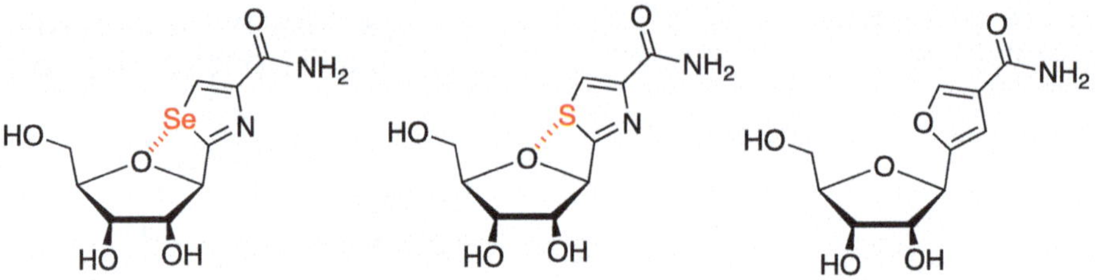

6 **7** **8**

Fig. 4.7 Structures of nucleoside analogues

Table 4.1 Structures and activities of Abl kinase inhibitors **9–15**[a]

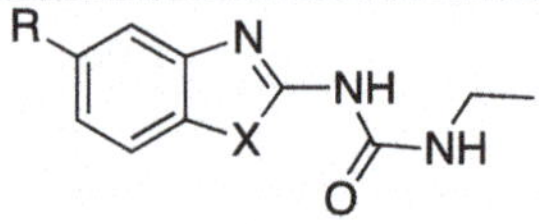

Cmp	R	X	IC_{50} (nM)
9	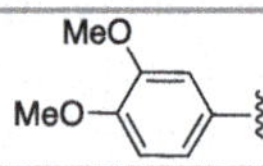	NH	110
10		S	1.99
11	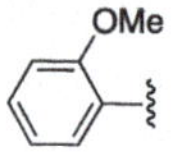	NH	18.8
12		S	0.06
13		O	40.8
14	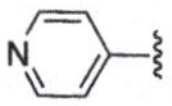	NH	70
15		S	1.12

[a]The data were taken from Ref. [49]

the σ-hole with the lobe projecting toward the ribose *O*-4 atom, which has the larger electrostatic potential. Interestingly, the ability to bind IMPDH well correlates with the possibility to form CB, with the selenium derivative being the most potent among the series (**6** > **7** > **8**). This indicated that this nonbonded interaction is pivotal for a proper enzymatic inhibition [48] (Fig. 4.7).

If compared to 1,4, intramolecular 1,5 *O*--*S* interactions are energetically more favorable because of the expanded geometry, that facilitates the overlap between the lone pair of electrons on oxygen and the σ-hole of sulfur. Besides that shown in Fig. 4.5, a further nice example is provided by the Abl kinases inhibitors **9–15**, showed in Table 4.1. In all the cases, the benzothiazoles are considerably more potent than the corresponding benzimidazoles **9**, **11** and **14** or the benzoxazole **13**, with a 50–600-fold variation [49, 50]. The reason behind the higher activity is that just sulfur is capable to establish CBs, reducing the freedom of the urea group thus facilitating the target recognition.

Both 1,4 and 1,5 *O*--S intramolecular interactions were found in the crystal structure of the antiparasitic nitazoxanide (compound **16**, Fig. 4.8). The endocyclic sulfur is coordinated by the nitro oxygen (1,4 interaction) and the amidic one (1,5 interaction). The interatomic distances are of 2.9 and 2.7 Å, respectively, both

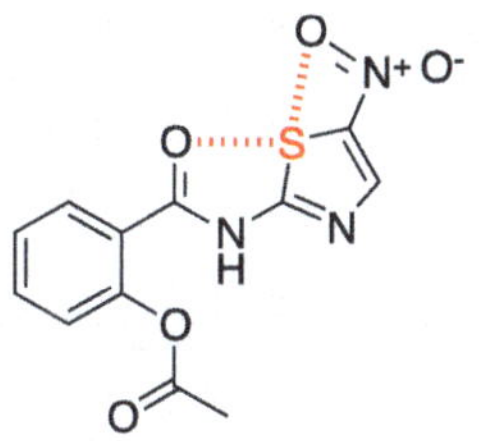

Fig. 4.8 Structure of nitazoxanide

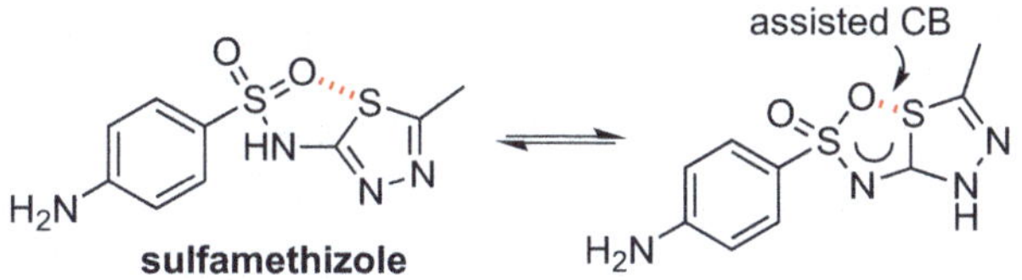

Fig. 4.9 Resonance-assisted CB in sulfamethizole

smaller than the sum of the van der Waals radii of the engaged atoms. This unusual bis-coordination is facilitated by the powerful electron-withdrawing effect of the NO_2 group, which increases the size of the σ-hole [51].

The CB contributes to the stabilization of several tautomeric forms of the dihydrofolate reductase inhibitor sulfamethizole (Fig. 4.9). In the solid state, a resonance-assisted *S--O* CB in a five-membered ring was found for the principal tautomer. The strength and the directionality of the CB can be manipulated by co-crystallization with *p*-aminobenzoic acid, protonation or deprotonation [52].

4.5 Chalcogen Bond in Organic Synthesis

Weak interactions are widely applied in synthesis, and the CB starts to be appreciated among the organic chemistry community whit a growing number of examples focused on the CB-assisted synthesis that have been continuously published. Among the first examples, it was reported that the *S--O* CB was important to rationalize the high diastereoselectivity observed after the reaction between the hydroxyl epoxides **17** and **18** (Scheme 4.1) with NaSPh in THF. A mixture of the corresponding γ-butyrolactones **19** (*syn–syn*) and **20** (*syn–anti*) were obtained in a 31:69 ratio, with none of the *anti–syn* or *anti–anti* isomers observed. The single-crystal X-ray structure of compound **19** showed that the interatomic distance between *S* and the hydroxyl *O* was of 2.9 Å, with the *S*–phenyl ring oriented diametrically away from the *OH*. This stereochemical arrangement would permit also a CB in which *S* is donating one of its lone pairs to the π* orbital of the adjacent carbonyl (blue arrow in Scheme 4.1). A retroaldol/aldol reaction (depicted in the lower panel of Scheme 4.1) served to rationalize both the equilibrium and the lack of formation of *anti–syn* and *anti–anti* diastereoisomers, in which the key intermediate (depicted in square

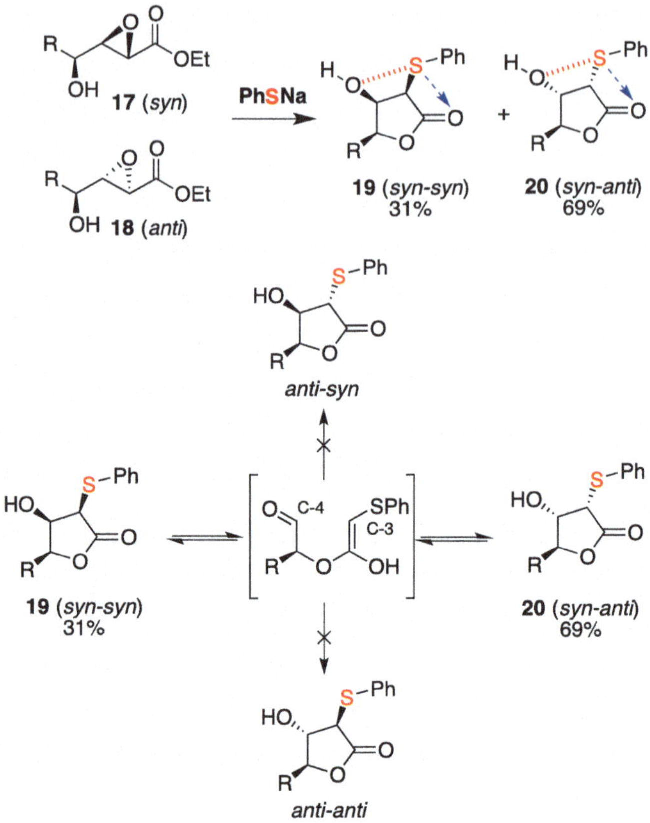

Scheme 4.1 Reaction between the hydroxyl epoxides and PhSNa

brackets) gives an aldol cyclization to furnish a thermodynamic mixture of lactones. The tendency of the hydroxyl group and the thiol to adopt a *syn* relationship could be explained by the *S--O* interaction in the enolate–aldehyde, which would orientate the aldehyde and the sulfurated enolate in the *syn* position [53, 54].

The importance of *Se*--X CB, where X is a lone pair-containing group, is particularly evident with electrophilic organoselenium reagents used in asymmetric syntheses. In particular, the electrophilic addition of chalcogenides to unsaturated substrates in the presence of external or internal nucleophiles is a well-known protocol to functionalize C-C double and triple bonds. The stereoselectivity is correlated to the intermediate formation of a chalcogeniranium ion, that drives a *trans* addition to the double or triple bonds enabling the installation of several functional groups such as alcohols, ethers, amides, azides, halides or the formation of heterocycles when the substrate is suitably functionalized with an internal nucleophile (Scheme 4.2) [55, 56].

As shown in the upper panel of Scheme 4.2, the electrophilic reagents are prepared starting from the corresponding diselenides, that are relatively easier to be

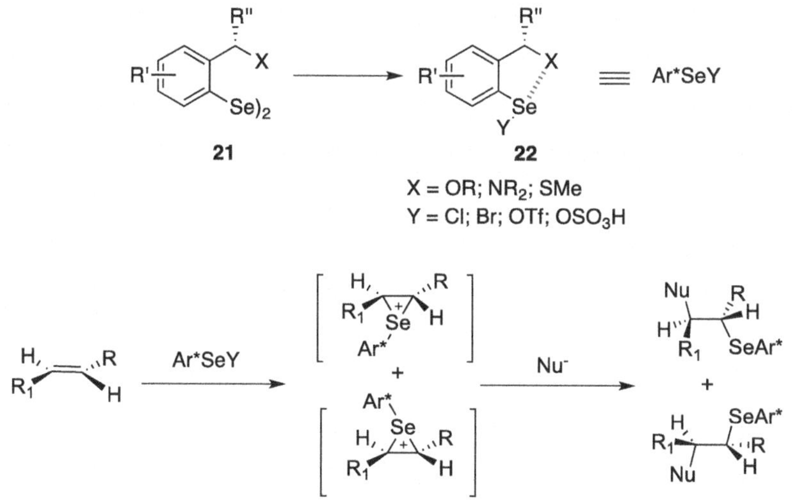

Scheme 4.2 Electrophilic selenofunctionalization of olefins

prepared and handled and can be conveniently converted into, besides electrophiles, also nucleophilic and radical reagents [57]. The aromatic diselenides depicted in Scheme 4.2, from a structural point of view, are characterized by a double substitution in their benzylic position. One substituent is responsible for the steric hindrance (**R″**), while the second one is a lone pair-containing group (**X**), which is the CB donor once the diselenide is converted into the electrophilic species.

The intramolecular CB between selenium and a close heteroatom is a very important factor, which is responsible for the efficiency of various chiral electrophilic selenium reagents in asymmetric syntheses reported so far. Such an interaction creates a highly stable five-membered cycle that forces the asymmetric carbon to approach the reaction centre during the addition, thus improving the chirality transfer. Several CB donors have been tested in the X position [58], but among all the groups, the best results were obtained with SMe. For this reason, diselenides precursors **23** and **24** (Scheme 4.3) will be taken as prototypical of the whole class and discussed here in more details.

The efficiency in stereoselective syntheses of the electrophilic reagent **25**, prepared by treating diselenide **23** with Br_2 and then with AgOTf, was tested by determining the diastereomeric ratios obtained in the selenomethoxylation and selenohydroxylation reactions of alkenes (Scheme 4.3), which afforded compounds **26** and **27**, respectively, as a mixture of the two possible diastereoisomers. The addition of **23** to an alkene gives rise to a mixture of two diastereomeric seleniranium intermediates (lower panel of Scheme 4.2), which are trapped by the nucleophile to afford a mixture of the two enantiomerically pure diastereomeric products **26** and **27** derived from the stereospecific *anti* addition process. Carrying out the selenomethoxylation and selenohydroxylation reactions at −78 °C and at 0 °C respectively, good yields and very good diastereoselectivities were obtained [59].

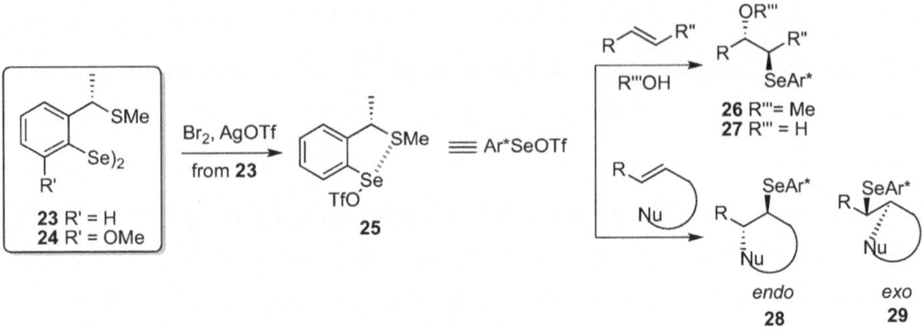

Scheme 4.3 Applications of the electrophilic reagent **25**

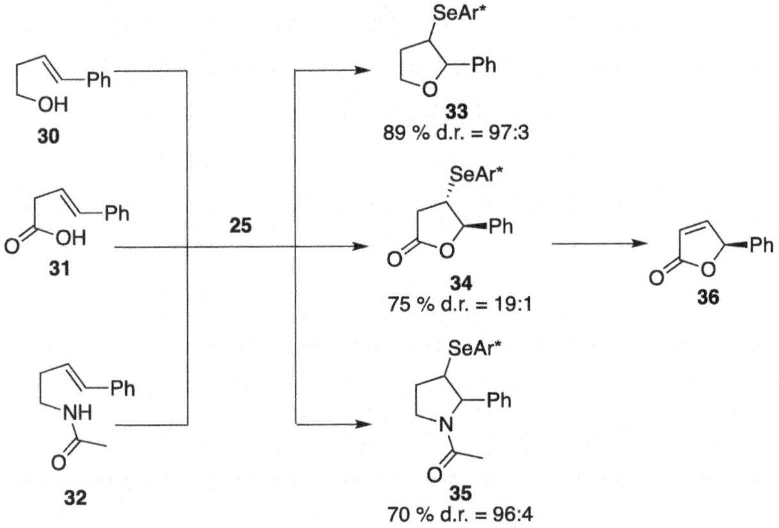

Scheme 4.4 Synthesis of heterocycles using the electrophilic reagent **25**

Successively, compound **25** was efficiently used to promote the ring-closure reaction of alkenes bearing a suitably positioned oxygen- or nitrogen-containing nucleophilic group. This is a versatile strategy to obtain enantiomerically enriched heterocyclic compounds, such as ethers, lactones, lactams and *N*-protected pyrrolidines. From a regioselectivity point of view, the reaction products can be the result of an *endo* (compound **28**) or an *exo* (compound **29**) attack based on the relative positions of the double bond and the nucleophilic group [60].

Alkenols (compound **30**, Scheme 4.4), alkenoic acids (**31**) and *N*-alkenyl acetamides (**32**) were converted into the corresponding heterocycles with good yields and high diastereoselectivities (Scheme 4.4) by using the electrophilic reagent **25**. For compound **34**, the absolute configuration was determined after its conversion into the known butenolide **36**, which was also investigated for its enantiomeric

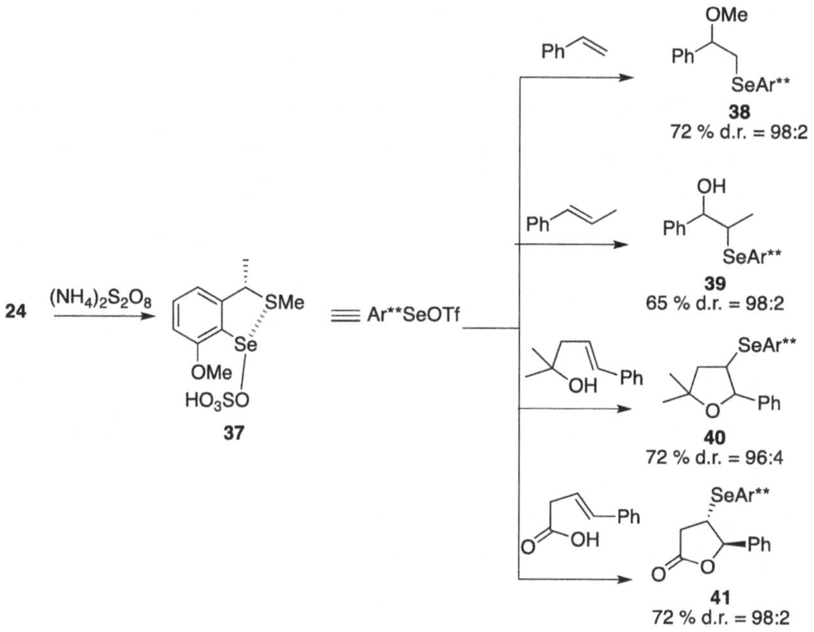

Scheme 4.5 Applications of the electrophilic reagent **37**

content, resulting, as expected, identical to the diastereomeric excess of the starting products [60].

Diselenide **24** was designed by the introduction of a sterically hindered substituent in the aromatic C-6 position, this modification would have improved the stereoselection capability of the resulting electrophilic specie. Once converted into the corresponding selenyl sulphate (compound **37**, Scheme 4.5), it was used to promote several electrophilic transformations of the C-C double bonds.

As expected, the introduction of the methoxy group in the *ortho* position respect to the electrophilic centre improved the outcome of the reactions, both in terms of yields and stereoselection [61]. Compound **24** was then used as electrophilic precursor in a series of asymmetric azidoselenylation of C-C double bonds. This is one among the very few examples through which the installation of these precious functional groups is possible [62]. The same diselenide was used in the first example of kinetic resolution performed using organoselenium compounds. The kinetic resolution is the process through which a racemic mixture can be resolved exploiting the different reaction rates of the two enantiomers with the same reagent [63].

As exemplified in Scheme 4.6, a racemic mixture of the alkenol **42** reacts with 0.5 equiv. of the optically active selenorganic compound **43**, leading to the quantitative formation of the product **44**, which comes from the preferential reaction with one enantiomer. At the end of the kinetic resolution, the stating material is recovered in its enantiomerically enriched form (compound **42a**). The preference for the *beta* enantiomer can be explained by the transient chalcogen bond established between

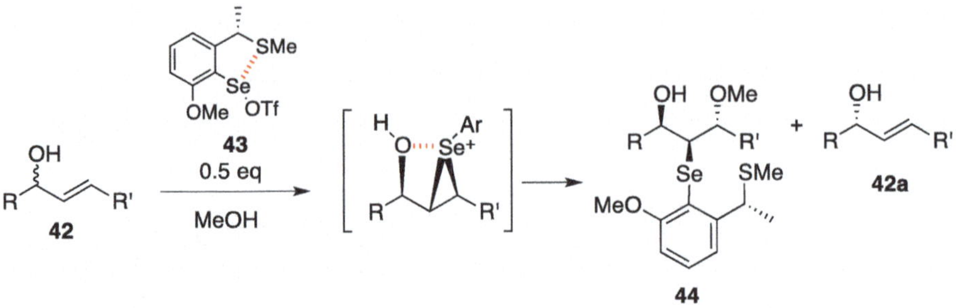

Scheme 4.6 Kinetic resolution using the electrophilic reagent **43**

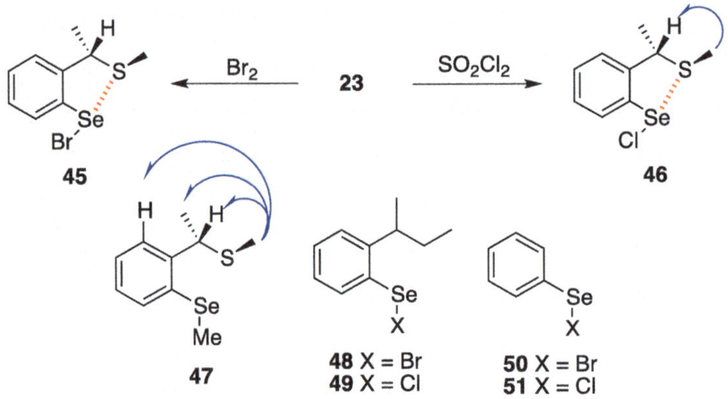

Scheme 4.7 Studies on intramolecular non-bonded interactions

the electron-poor seleniranium ion and the hydroxyl group present in the substrate [64]. Very recently, a similar approach was applied for the kinetic resolution of 2-methoxycarbonylalk-3-enols [65].

The existence of CB for compounds **45** and **46** (Scheme 4.7) was demonstrated through X-ray analysis and NMR techniques. They were prepared starting from diselenide **23** by mean of a bromination and a chlorination reaction, respectively. The X-ray analysis revealed that compounds **45** and **46** are isostructural with the selenium atom, having a T-shaped coordination geometry with a bond angle of 178.9° for **45** and 177.7° for **46**. The distance between *Se* and *S* atoms [2.497 Å for **45** and 2.344 Å for **46**] is significantly shorter than the sum of their van der Waals radii (3.7 Å), and this demonstrates that an intramolecular interaction between selenium and sulfur actually exists. The shorter distance observed in compound **46** compared with that of the bromide analogue seems to indicate a stronger interaction when the counter ion is chloride, and this observation is consistent with the calculated covalency factor χ for bromide (0.826) and chloride (0.902) derivatives.

The Overhauser dipolar correlations for arylselenenyl chloride **46** were measured and compared with those obtained for the corresponding arylmethyl selenide **47**. As indicated in Scheme 4.7, compound **46** presents a greater conformational rigidity than that observed in **46**. Indeed, while clear NOE correlations were observed for the SMe group (blue arrows) in **47**, the same were lacking in **46**, with the exception of that with the beta-hydrogen in the asymmetric carbon, which indicates that the CB exists not only in the crystal form but also in solution. The existence of CB was also demonstrated by its effect on proton and carbon-13 resonances of the far methyl and methine groups. In order to evaluate the influence of the *Se--S* interaction on the ^{77}Se NMR chemical shift values, *ortho* alkyl-substituted selenyl halides derivatives **48** and **49**, in which the sulfur is replaced by a carbon atom, were synthesized. The steric effect results in an upfield shift of the ^{77}Se NMR signals of bromide **48** (δ = 832 ppm) and chloride **49** (δ = 1004 ppm) with respect to the unsubstituted PhSeBr **50** (δ = 867 ppm) and PhSeCl **51** (δ = 1044 ppm). The presence of CB determines a further upfield shift (δ = 750 ppm for compound **45** and 797 ppm for compound **46**) [66]. This is in agreement with the trend observed by Tomoda in the case of the *Se--O* interaction [67].

A 1,5 *N--Se* interaction was found in the single-crystal X-ray structure of bis{[2-(*N*,*N*-dimethylamino)methyl]phenyl}tetraselenide (compound **52**), which is per se a quite unusual functional group due to its instability. The presence of such CB makes the isolation of this elusive Se_4 arrangement possible. The reason behind the formation of the organotetraselenide is the presence of 2 five-membered cycles with the assistance of intramolecular chalcogen bonding [68].

Being oxygen the most electronegative and the least polarizable among chalcogens, CBs with oxygen as acceptor are rare. Among the most recent examples, that reported in 2012 by Zhu and coworkers is worth to be mentioned. In the hypervalent iodine derivative **53**, the interatomic distance between the carbonyl and the ester oxygens (2.734 Å) was found to be inferior to the sum of the van der Waals radii, thus representing a clear nonbonding interaction [69] (Fig. 4.10).

Despite the number of reported examples, among chalcogens, tellurium is thought to be the best CB acceptor. An interesting example of CB in which *Te* is engaged was reported in 2009 in the contest of the synthesis of 3,4 disubstituted tellurophenes. The ring-closure addition–elimination reaction of 2,3-dimethoxybuta-1,3-diene with $TeCl_4$ in the presence of $(CH_3Si)_2$ and CH_3COONa yields tellurophene stabilized by strong intermolecular CBs (Scheme 4.8). The metallic character of tellurium accounted for the formation of the *Te--Te* CB leading to the formation of the six-membered ring in 3,4-methoxytellurophene [70].

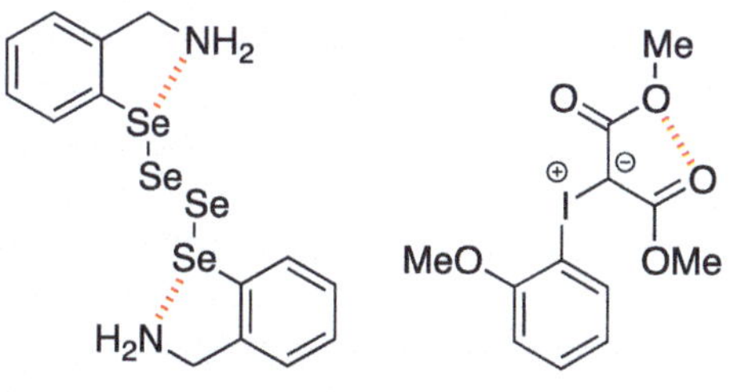

52 53

Fig. 4.10 Structures of compounds **52** and **53**

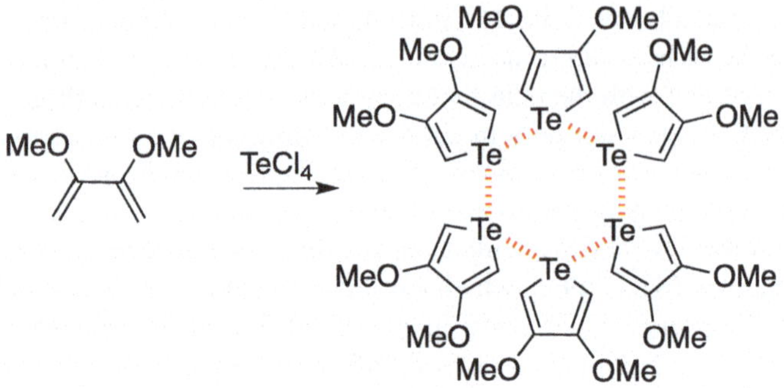

Scheme 4.8 Synthesis of CB-stabilized tellurophenes

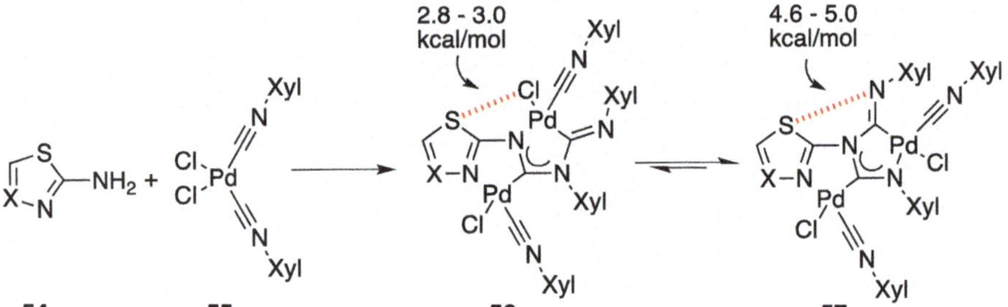

Scheme 4.9 Synthesis of binuclear diaminocarbene-Pd(II) complexes

In the research field of coordination compounds, the presence of CBs drives the molecular assembly process and influences the properties of the target molecules. An interesting example is found in the synthesis of binuclear diaminocarbene-Pd(II) complexes, where two distinct types of CBs modulate the regioisomerization of the target compound. In particular, the reaction between 1,3-thiazol-2-amines (compound **54**) and palladium isocyanide **55** yields two isomers; one in which a *S--N* (compound **57**, thermodynamically controlled) CB is found and the other bearing an intramolecular *S--Cl* (compound **56**, kinetically controlled isomer) (Scheme 4.9). Carrying out the reaction at room temperature, a mixture in which compounds **56** were in a larger amount over the thermodynamic regioisomer was obtained. The same reaction under refluxing conditions gave the opposite ratio [71]. The X-ray analysis of compounds **56** and **57** revealed the presence of the two CBs that, on the basis of the Atoms in Molecules analysis [72], were ranked in terms of strength. Both interactions showed energies typical of CB, but the *S--N* contacts (4.9–6.0 kcal/mol in the solid state and 4.6–5.3 kcal/mol in $CHCl_3$) were stronger than the *S--Cl* (3.1–3.2 kcal/mol in the solid state and 2.8–3.0 kcal/mol in $CHCl_3$) ones. As expected, the difference in energies of *S--N* and *S--Cl* contacts correlates well with the relative stability of these isomers [71].

Besides organic chemistry, there are several examples in which CBs were exploited in inorganic synthesis, especially in the preparation of coordination compounds [6]. Similar to the example reported in Scheme 4.9, some palladium-containing supramolecular structures stabilized by *Te--O* CB have been reported quite recently [73]. In general, the metal coordination increases the electrophilicity of the chalcogen atom of a ligand, making it more prone to making CBs [6].

4.6 Chalcogen Bond in Catalysis

Noncovalent interactions play an important role in many catalytic processes offering stabilization to intermediates and providing, in the case of asymmetric synthesis, high enantioselectivity and diastereoselectivity [74]. Although these interactions are individually weak, their orchestrated action can provide a powerful tool to reach high yields while controlling selectivity. CB-based catalysts are relatively rare if compared to others whose activity is based on conventional interactions such as π-stacking [75], hydrogen [76], and halogen [77] bonding. For this reason, it is possible to state that the CB-aided catalysis is still in its infancy; nevertheless, it is considered a fascinating research field.

To the best of our knowledge, among the few examples reported in the literature, the one-pot selenenylation–deselenenylation sequences performed with catalytic amounts of the optically active selenyl sulfate **37** is the first example. As shown in Scheme 4.10, the sequence is carried out with compound **37**, generated from diselenide **24** through the reaction with ammonium persulfate, as the electrophilic reagent for the initial addition to an alkene (compound **58**). The resulting selenide

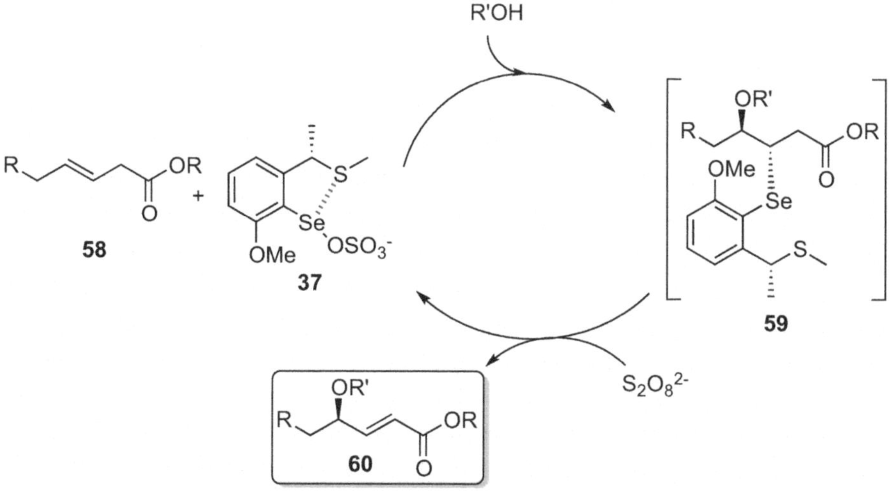

Scheme 4.10 One-pot selenenylation-deselenenylation of olefins

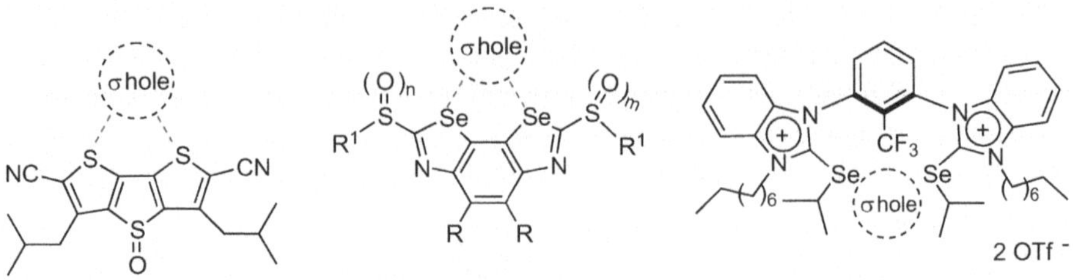

61
DTT

62
BDS

63a

Fig. 4.11 Sulfur- and selenium-containing chalcogen-bonding catalysts

59 is then oxidized by the excess of persulfate, affording the deselenenylated compound through an elimination reaction and, at the same time, regenerating the selenenyl sulfate, which is again ready to catalyze the same reaction (see Chap. 1 for a detailed discussion on *Se*-catalyzed reactions) [61].

Actually, in this case the chalcogen bond is required for the stereoselection process rather than the catalysis itself. Conversely, Matile and coworkers described the very first catalysts operating via a CB, which were based on dithieno[3,2-*b*:2′,3′-*d*]-thiophenes (DTTs) such as compound **61** [78], i.e., sulfur chalcogen bond donors. Few months later, the same group described the synthesis and application of a new class of engineered benzodiselenazoles (BDS) **62** that fulfill, according to the authors, all the requirements for a high-precision chalcogen-bonding catalysis: conformationally immobilized σ-holes on highly electron-deficient *Se* atoms in a neutral platform (Fig. 4.11) [79].

Twelve differently substituted BDSs **62** were synthesized and evaluated for their activity in promoting the hydrogenation of quinoline **64** with the Hantzsch ester **65**. Catalyst **63a** (R^1 = 4-^{t}BuC_6H_4), with strong electron-withdrawing groups; two sulfones (n = m = 2) and two cyano groups (R = CN), was the more active, affording excellent yields of the respective 1,2,3,4-tetrahydroquinolines **66** after 24–48 h of reaction (Scheme 4.11).

The presence of four electron-acceptor groups in the scaffold causes a deepening in the σ-holes of the *Se* and allows stronger R-*Se*--*N* chalcogen bonds between **62a** and the anion **64** generated after attack of a hydride from **65**. The authors reasoned that this interaction in the focal σ-holes of **62a** enables the transition state stabilization due to the formation of a five-membered ring, resulting in rate acceleration of the transfer hydrogenation or, in other words, the chalcogen bond catalysis (Fig. 4.12).

Huber and coworkers designed new bis(2-selanylbenzimidazolium) derivatives **63** and tested them as chalcogen bond donors in the solvolysis of benzhydryl bromide **67** in wet acetonitrile (Scheme 4.12) [80]. Among the prepared selenides, the best result was obtained using a stoichiometric amount of catalyst **63a**, which caused a 20–30-fold increase in the reaction rate vs. the background solvolysis, giving the expected deuterated amide **68** in 64% yield after 24 h at room temperature (Scheme 4.12). This performance was superior to that of the close related

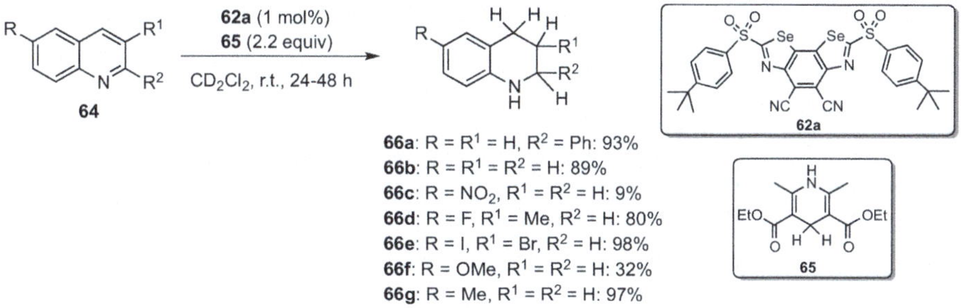

Scheme 4.11 Hydrogenation of quinoline with the Hantzsch ester

Fig. 4.12 Mechanism of the chalcogen-bonding catalysis using **62a**

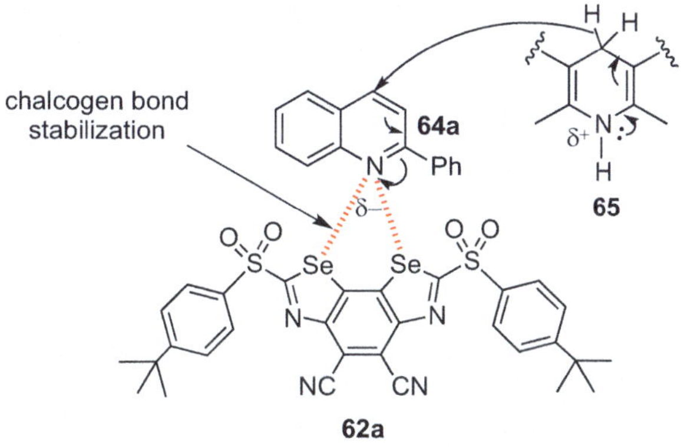

brominated halogen bond donor, but inferior to the respective iodide. Several experiments were performed to collect evidences that the rate acceleration was due to chalcogen bonding since the non-selenylated reference compound was inactive. Despite a stoichiometric amount of **63a** was used (so here a "*Se*-promoted," not "*Se*-catalyzed" reaction should be a more suitable denomination), the outcomes from Huber group are the first evidence of an intermolecular chalcogen bond C-Br activation and can open new windows for the discovery of catalytic versions for this approach.

A further interesting example of CB-aided catalysis is reported in Scheme 4.13, where a chiral isothiourea catalyst (compound **69**) yields an α,β-unsaturated acylammonium intermediate which undergoes two sequential chemoselective nucleophilic attacks by 2-aminothiophenols, producing 1,5-benzothiazepines. On the basis of the experimental results, the authors proposed the reaction mechanism outlined in Scheme 4.13. Starting from α,β-unsaturated (*E*)-anhydride, the acylammonium intermediate **70** is formed with the carbonyl group fixed in its position through a *S--O* CB. The 2-aminothiophenol can then approach from the opposite side of the phenyl group on the catalyst for steric reasons. Subsequently, a thia-Michael addition followed by *N*-acylation affords the desired cycloadduct (R)-**71** [81].

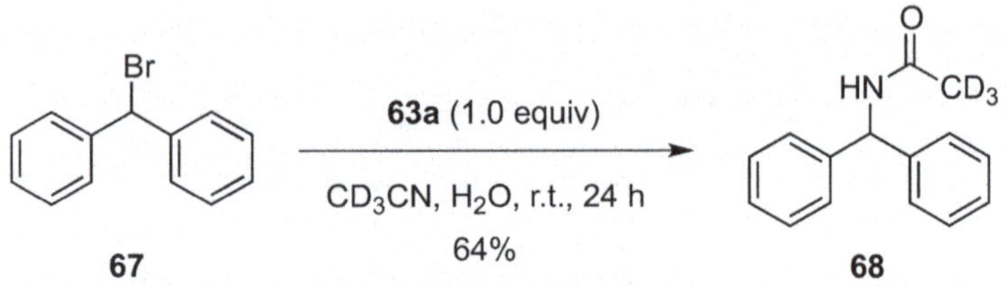

Scheme 4.12 Solvolysis of benzhydryl bromide

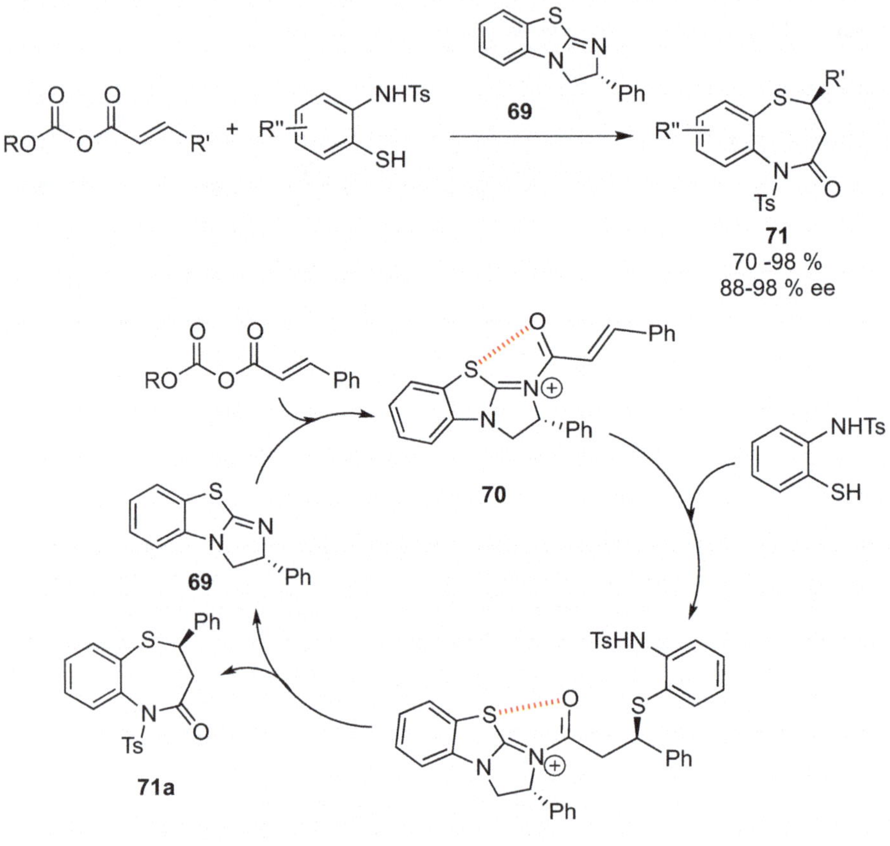

Scheme 4.13 CB-assisted synthesis of 1,5-benzothiazepines

4.7 Chalcogen Bond in the Chemistry of Materials

As all the noncovalent interactions do, chalcogen bonding can control the intermolecular and intramolecular arrangement, both in the solid state and in solution. For this reason, non-bonded interactions are very important for the successful design of materials with tunable properties. If compared with hydrogen and halogen bonding,

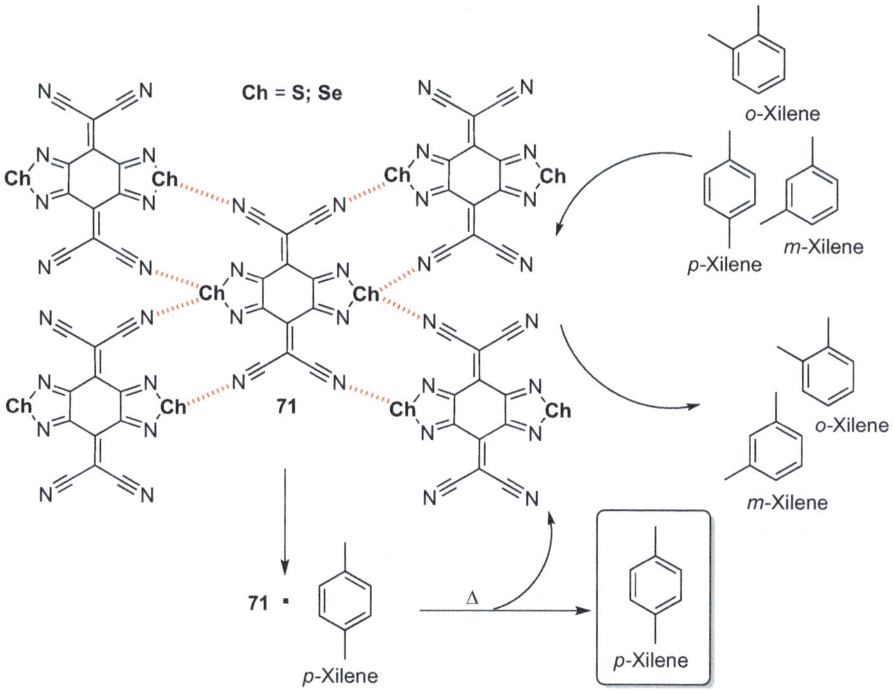

Fig. 4.13 Separation of xylene isomers through a CB aided process

a few examples of the application of CBs for the design of new materials have been reported; some of them will be analyzed in the present section.

Among the pioneering examples of materials, which owe their function to the presence of CB, the lattice med of tetracyanoquinodimethanes fused with thiadiazole and selenadiazole rings, reported in 1992 by Miyashi and coworkers, is of worth mentioning. The intermolecular arrangement, in which it is possible to recognize a dense network of Ch--*N* CBs, creates a material (Fig. 4.13) able to separate very similar isomers by selective complexation and thermal decomplexation. During the crystallization process in the presence of all of the xylene isomers, the authors realized that compound **71** was able to form a 1:1 complex with *p*-xylene leaving out *o*- and *m*-xylene. In addition, the thermal decomposition allowed to obtain *p*-xylene in its pure form while fully recovering **71**. The same was done using a mixture of 2,6- and 2,7-dimethylnaphthalenes (DMNs), and the material was able to discriminate the 2,6-DMN with an efficiency comparable to that observed in the case of xylenes [82].

The development for receptors able to selectively capture small and potentially toxic ions is a fertile research field, especially for the investigation focused on the synthesis and anion-binding properties of polydentate Lewis acids. In 2010, Gabbai and coworkers synthesized a bidentate Lewis acid that contains a boryl and a chalconium moiety (compounds **72**, Fig. 4.14a). Anion-complexation studies indicate

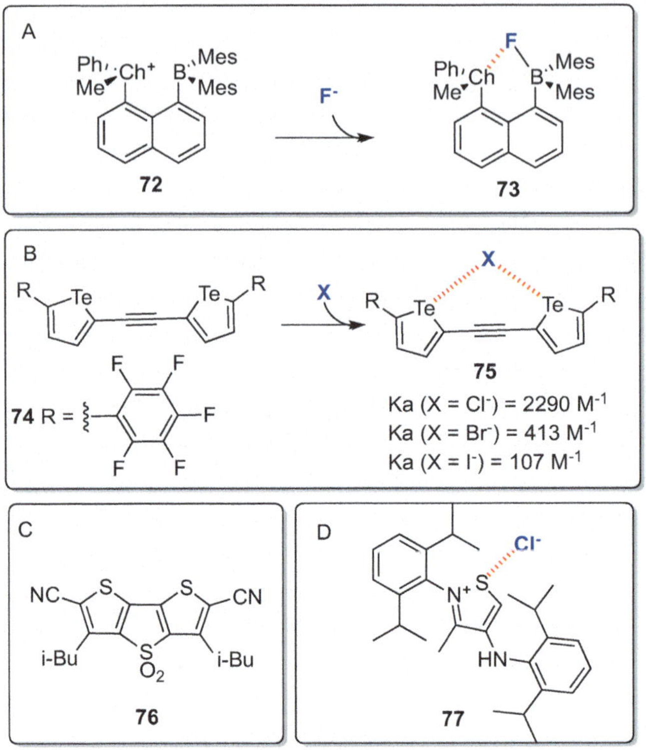

Fig. 4.14 Anion-binding molecules

that this borane displayed a high affinity for fluoride in methanol, whit both sulfur and tellurium as chalconium counterpart. The presence of *Te* in the receptor increases its affinity for fluoride because of the larger size of *Te* that allows for a reduction of intraligand repulsions, leading to a more favorable chalcogen bonding with the fluoride analyte [83].

Very recently, electron-deficient 2,5-diaryltellurophene (compound **74**, Fig. 4.14b) displayed appreciable anion affinity in an organic solvent. Interestingly, the affinity was improved using a bis(tellurophene) receptor in which the chalcogen-based heterocycles are joined by an ethynylene linker. Compound **74** displayed different affinities according to the anion. The complex with *Cl* was modelled via DFT calculation and the minimum-energy geometry is stabilized by a chalcogen bonding interaction having a *Te--Cl* distance of 3.07 Å and a 168° *Cl--Te–C* angle [84].

The anion binding ability of chalcogen containing molecules was exploited to generate anion transporters. In particular, the dithieno[3,2-*b*;2′,3′-*d*]thiophenes (DTTs) were identified as ideal structures to bind anions and transport them across lipid bilayers. Both the binding and transport capabilities are improved by increasing the depth of the σ-holes in the chalcogen atoms. Transport activities were assessed in large unilamellar vesicles composed by phosphatidylcholine and loaded with 8-hydroxy-1,3,6-pyrenetrisulfonate (HPTS), which is a pH sensitive fluorescent probe. A pH gradient was applied first, and then the transporters were added and their ability to accelerate the dissipation of the pH gradient was measured and expressed in terms of half maximal effective concentration (EC_{50}). The best in class was compound **76** (Fig. 4.14c), which showed a EC_{50} = 1.9 μM. This activity was in agreement with the affinity for chloride (K_D = 1.13 mM) and the highest theoretical binding energy $E_{int} = -34.6$ kcal/mol, which is the deepest σ-hole [85].

In the last anion-binding molecule (compound **77**), it is possible to observe a cooperation of a negative charge-assisted chalcogen bonding with ionic interactions. In the thiazole compound, the *S--Cl* distance is of 2.848 Å, which is significantly lower than 3.55 Å, the sum of the *S* and *Cl* van der Waals radii. In addition, the intramolecular distance between the chloride anion and the N^+ of the thiazole ring is 4.553 Å, which fits well into the range of ionic interaction distances. In this example, the cooperation of multiple noncovalent interactions contributes to the stabilization of the organic salt in the solid state [86].

References

1. Scheiner S (2003) Nonbonded interactions. In: Bultinck P, De Winter H, Langenaeker W, Tollenaere JP (eds) Computational medicinal chemistry for drug discovery. CRC Press, New York, pp 235–257
2. DiLabio GA, Otero-de-la-Roza A (2016) Noncovalent interactions in density functional theory. In: Parril AL, Lipkowitz KB (eds) Reviews in computational chemistry, vol 29. Wiley, Hoboken, pp 1–97
3. Breugst M, von der Heiden D, Schmauck J (2017) Novel noncovalent interactions in catalysis: a focus on halogen, chalcogen, and anion-π bonding. Synthesis 49:3224–3236
4. Metrangolo P, Resnati G (2001) Halogen bonding: a paradigm in supramolecular chemistry. Chemistry 7:2511–2519
5. Scheiner S (2013) The pnicogen bond: its relation to hydrogen, halogen, and other noncovalent bonds. Acc Chem Res 46:280–288
6. Mahmudov KT, Kopylovich MN, Guedes da Silva MFC, Pombeiro AJL (2017) Non-covalent interactions in the synthesis of coordination compounds: recent advances. Coord Chem Rev 345:54–72
7. Bauzá A, Mooibroek TJ, Frontera A (2013) Tetrel-bonding interaction: rediscovered supramolecular force? Angew Chem Int Ed 52:12317–12321
8. Bauzá A, Frontera A (2015) Aerogen bonding interaction: a new supramolecular force? Angew Chem Int Ed 54:7340–7343
9. Brookhart M, Green MLH (1983) Carbon-hydrogen-transition metal bonds. J Organomet Chem 250:395–408
10. Nagao Y (2013) Chemical pharma-sciences that incorporate non-covalent bonded interactions. Heterocycles 87:1–29
11. Clark T, Hennemann M, Murray JS, Politzer P (2007) Halogen bonding: the σ-hole. J Mol Model 13:291–296
12. Brinck T, Murray JS, Politzer P (1992) Surface electrostatic potentials of halogenated methanes as indicators of directional intermolecular interactions. Int J Quantum Chem 44:57–64
13. Politzer P, Murray JS, Clark T, Resnati G (2017) The σ-hole revisited. Phys Chem Chem Phys 19:32166–32178

14. Politzer P, Lane P, Concha MC, Ma Y, Murray JS (2007) An overview of halogen bonding. J Mol Model 13:305–311
15. Pecina A, Lepšík M, Hnyk D, Hobza P, Fanfrlík J (2015) Chalcogen and pnicogen bonds in complexes of neutral icosahedral and bicapped square-antiprismatic heteroboranes. J Phys Chem A 119:1388–1395
16. Scheiner S (2013) Detailed comparison of the pnicogen bond with chalcogen, halogen, and hydrogen bonds. Int J Quantum Chem 113:1609–1620
17. Mahmudov KT, Kopylovich MN, Guedes da Silva MFC, Pombeiro AJL (2017) Chalcogen bonding in synthesis, catalysis and design of materials. Dalton Trans 46:10121–10138
18. Wang H, Wang W, Jin WJ (2016) σ-Hole bond vs π-hole bond: a comparison based on halogen bond. Chem Rev 116:5072–5104
19. Vargas R, Garza J, Dixon DA, Hay BP (2000) How strong is the C α –H···OC hydrogen bond? J Am Chem Soc 122:4750–4755
20. Gallivan JP, Dougherty DA (2000) A computational study of cation–π interactions vs salt bridges in aqueous media: implications for protein engineering. J Am Chem Soc 122:870–874
21. Brandl M, Weiss MS, Jabs A, Sühnel J, Hilgenfeld R (2001) C-H···π-interactions in proteins. J Mol Biol 307:357–377
22. Iwaoka M, Isozumi N (2012) Hypervalent nonbonded interactions of a divalent sulfur atom. Implications in protein architecture and the functions. Molecules 17:7266–7283
23. Row TNG, Parthasarathy R (1981) Directional preferences of nonbonded atomic contacts with divalent sulfur in terms of its orbital orientations. 2. Sulfur.cntdot..cntdot..cntdot.sulfur interactions and nonspherical shape of sulfur in crystals. J Am Chem Soc 103:477–479
24. Iwaoka M, Isozumi N (2006) Possible roles of S···O and S···N interactions in the functions and evolution of phospholipase A2. Biophysics (Oxf) 2:23–34
25. Moroder L (2005) Isosteric replacement of sulfur with other chalcogens in peptides and proteins. J Pept Sci 11:187–214
26. Iwaoka M, Babe N (2015) Mining and structural characterization of S···X chalcogen bonds in protein database. Phosphorus Sulfur Silicon Relat Elem 190:1257–1264
27. Nachman J, Miller M, Gilliland GL, Carty R, Pincus M, Wlodawer A (1990) Crystal structure of two covalent nucleoside derivatives of ribonuclease A. Biochemistry 29:928–937
28. Taylor JC, Markham GD (1999) The bifunctional active site of S-adenosylmethionine synthetase. J Biol Chem 274:32909–32914
29. Brandt W, Golbraikh A, Täger M, Lendeckel U (1999) A molecular mechanism for the cleavage of a disulfide bond as the primary function of agonist binding to G-protein-coupled receptors based on theoretical calculations supported by experiments. Eur J Biochem 261:89–97
30. Singh AK, Singh N, Sharma S, Shin K, Takase M, Kaur P, Srinivasan A, Singh TP (2009) Inhibition of lactoperoxidase by its own catalytic product: crystal structure of the hypothiocyanate-inhibited bovine lactoperoxidase at 2.3-Å resolution. Biophys J 96:646–654
31. Nakamura T, Yamamoto T, Abe M, Matsumura H, Hagihara Y, Goto T, Yamaguchi T, Inoue T (2008) Oxidation of archaeal peroxiredoxin involves a hypervalent sulfur intermediate. Proc Natl Acad Sci U S A 105:6238–6242
32. Beno BR, Yeung KS, Bartberger MD, Pennington LD, Meanwell NA (2015) A survey of the role of noncovalent sulfur interactions in drug design. J Med Chem 58:4383–4438
33. Achibat H, AlOmari NA, Messina F, Sancineto L, Khouili M, Santi C (2015) Organoselenium compounds as phytochemicals from the natural kingdom. Nat Prod Commun 10:1885–1892
34. Sancineto L, Mariotti A, Bagnoli L, Marini F, Desantis J, Iraci N, Santi C, Pannecouque C, Tabarrini O (2015) Design and synthesis of DiselenoBisBenzamides (DISeBAs) as nucleocapsid protein 7 (NCp7) inhibitors with anti-HIV activity. J Med Chem 58:9601–9614
35. Sancineto L, Piccioni M, De Marco S, Pagiotti R, Nascimento V, Braga AL, Santi C, Pietrella D (2016) Diphenyl diselenide derivatives inhibit microbial biofilm formation involved in wound infection. BMC Microbiol 16:220
36. Bartolini D, Sancineto L, de Bem AF, Tew KD, Santi C, Radi R, Toquato P, Galli F (2017) Selenocompounds in cancer therapy: an overview. Adv Cancer Res 136:259–302

37. Santi C, Tidei C, Scalera C, Piroddi M, Galli F (2013) Selenium containing compounds from poison to drug candidates: a review on the GPx-like activity. Curr Chem Biol 7:25–36
38. Pacuła AJ, Mangiavacchi F, Sancineto L, Lenardão EJ, Ścianowski J, Santi C (2015) An update on "Selenium containing compounds from poison to drug candidates: a review on the GPx-like activity". Curr Chem Biol 9:97–112
39. Choi-Sledeski YM, Kearney R, Poli G, Pauls H, Gardner C, Gong Y, Becker M, Davis R, Spada A, Liang G, Chu V, Brown K, Collussi D, Leadley R, Rebello S, Moxey P, Morgan S, Bentley R, Kasiewski C, Maignan S, Guilloteau JP, Mikol V (2003) Discovery of an orally efficacious inhibitor of coagulation factor Xa which incorporates a neutral P 1 ligand. J Med Chem 46:681–684
40. Schärfer C, Schulz-Gasch T, Ehrlich HC, Guba W, Rarey M, Stahl M (2013) Torsion angle preferences in druglike chemical space: a comprehensive guide. J Med Chem 56:2016–2028
41. Tajima H, Honda T, Kawashima K, Sasabuchi Y, Yamamoto M, Ban M, Okamoto K, Inoue K, Inaba T, Takeno Y, Aono H (2010) Pyridylmethylthio derivatives as VEGF inhibitors. Part 1. Bioorg Med Chem Lett 20:7234–7238
42. Tajima H, Honda T, Kawashima K, Sasabuchi Y, Yamamoto M, Ban M, Okamoto K, Inoue K, Inaba T, Takeno Y, Tsuboi T, Tonouchi A, Aono H (2011) Pyridylmethylthio derivatives as VEGF inhibitors: Part 2. Bioorg Med Chem Lett 21:1232–1235
43. Lin S, Wrobleski ST, Hynes J, Pitt S, Zhang R, Fan Y, Doweyko AM, Kish KF, Sack JS, Malley MF, Kiefer SE, Newitt JA, McKinnon M, Trzaskos J, Barrish JC, Dodd JH, Schieven GL, Leftheris K (2010) Utilization of a nitrogen–sulfur nonbonding interaction in the design of new 2-aminothiazol-5-yl-pyrimidines as p38α MAP kinase inhibitors. Bioorg Med Chem Lett 20:5864–5868
44. Lee JC, Laydon JT, McDonnell PC, Gallagher TF, Kumar S, Green D, McNulty D, Blumenthal MJ, Keys JR, Land vatter SW, Strickler JE, McLaughlin MM, Siemens IR, Fisher SM, Livi GP, White JR, Adams JL, Young PR (1994) A protein kinase involved in the regulation of inflammatory cytokine biosynthesis. Nature 372:739–746
45. Dominguez C, Powers DA, Tamayo N (2005) p38 MAP kinase inhibitors: many are made, but few are chosen. Curr Opin Drug Discov Dev 8:421–430
46. Burling FT, Goldstein BM (1992) Computational studies of nonbonded sulfur-oxygen and selenium-oxygen interactions in the thiazole and selenazole nucleosides. J Am Chem Soc 114:2313–2320
47. Goldstein BM, Takusagawa F, Berman HW, Srivastava PC, Robins RK (1985) Structural studies of a new antitumor and antiviral agent: selenazofurin and its .alpha. anomer. J Am Chem Soc 107:1394–1400
48. Gebeyehu G, Marquez VE, Van Cott A, Cooney DA, Kelley JA, Jayaram HN, Ahluwalia GS, Dion RL, Wilson YA, Johns DG (1985) Ribavirin, tiazofurin, and selenazofurin: mononucleotides and nicotinamide adenine dinucleotide analogs. Synthesis, structure, and interactions with IMP dehydrogenase. J Med Chem 28:99–105
49. Park H, Hong S, Kim J, Hong S (2013) Discovery of picomolar ABL kinase inhibitors equipotent for wild type and T315I mutant via structure-based de novo design. J Am Chem Soc 135:8227–8237
50. Hong S, Kim J, Yun SM, Lee H, Park Y, Hong SS, Hong S (2013) Discovery of new benzothiazole-based inhibitors of breakpoint cluster region-Abelson kinase including the T315I mutant. J Med Chem 56:3531–3545
51. Félix-Sonda BC, Rivera-Islas J, Herrera-Ruiz D, Morales-Rojas H, Höpfl H (2014) Nitazoxanide cocrystals in combination with succinic, glutaric, and 2,5-dihydroxybenzoic acid. Cryst Growth Des 14:1086–1102
52. Thomas SP, Veccham SPKP, Farrugia LJ, Guru Row TN (2015) "Conformational simulation" of sulfamethizole by molecular complexation and insights from charge density analysis: role of intramolecular S···O chalcogen bonding. Cryst Growth Des 15:2110–2118

53. Rodríguez S, Kneeteman M, Izquierdo J, López I, González FV, Peris G (2006) Diastereoselective synthesis of γ-hydroxy A,β-epoxyesters and their conversion into β-hydroxy α-sulfenyl γ-butyrolactones. Tetrahedron 62:11112–11123
54. González FV, Jain A, Rodríguez S, Sáez JA, Vicent C, Peris G (2010) Stereoisomerization of β-hydroxy-α-sulfenyl-γ-butyrolactones controlled by two concomitant 1,4-type nonbonded sulfur−oxygen interactions as analyzed by X-ray crystallography. J Org Chem 75:5888–5894
55. Wirth T (ed) (2000) Organoselenium chemistry: modern developments in organic synthesis, vol 208. Springer, Berlin
56. Wirth T (1999) Chiral selenium compounds in organic synthesis. Tetrahedron 55:1–28
57. Santi C, Tomassini C, Sancineto L (2017) Organic diselenides: versatile reagents, precursors, and intriguing biologically active compounds. Chimia 71:592–595
58. Wirth T (2000) Organoselenium chemistry in stereoselective reactions. Angew Chem 39:3740–3749
59. Tiecco M, Testaferri L, Bagnoli L, Marini F, Temperini A, Tomassini C, Santi C (2000) Efficient asymmetric selenomethoxylation and selenohydroxylation of alkenes with a new sulfur containing chiral diselenide. Tetrahedron Lett 41:3241–3245
60. Tiecco M, Testaferri L, Marini F, Sternativo S, Bagnoli L, Santi C, Temperini A (2001) Sulfur-containing diselenide as an efficient chiral reagent in asymmetric selenocyclization reactions. Tetrahedron Asymmetry 12:1493–1502
61. Tiecco M, Testaferri L, Santi C, Tomassini C, Marini F, Bagnoli L, Temperini A (2002) Preparation of a new chiral non-racemic sulfur-containing diselenide and applications in asymmetric synthesis. Chem Eur J 8:1118–1124
62. Tiecco M, Testaferri L, Santi C, Tomassini C, Marini F, Bagnoli L, Temperini A (2003) Asymmetric azidoselenenylation of alkenes: a key step for the synthesis of enantiomerically enriched nitrogen-containing compounds. Angew Chem Int Ed 42:3131–3133
63. Robinson DEJE, Bull SD (2003) Kinetic resolution strategies using non-enzymatic catalysts. Tetrahedron Asymmetry 14:1407–1446
64. Tiecco M, Testaferri L, Santi C, Tomassini C, Bonini R, Marini F, Bagnoli L, Temperini A (2004) Chiral electrophilic selenium reagent to promote the kinetic resolution of racemic allylic alcohols. Org Lett 6:4751–4753
65. Tomassini C, Sarra FD, Monti B, Sancineto L, Bagnoli L, Marini F, Santi C (2017) Kinetic resolution of 2-methoxycarbonylalk-3-enols through a stereoselective cyclofunctionalization promoted by an enantiomerically pure electrophilic selenium reagent. ARKIVOC 2:303–312
66. Tiecco M, Testaferri L, Santi C, Tomassini C, Santoro S, Marini F, Bagnoli L, Temperini A, Costantino F (2006) Intramolecular nonbonding interactions between selenium and sulfur – spectroscopic evidence and importance in asymmetric synthesis. Eur J Org Chem 2006:4867–4873
67. Iwaoka M, Komatsu H, Katsuda T, Tomoda S (2004) Nature of nonbonded Se···O interactions characterized by 17 O NMR spectroscopy and NBO and AIM analyses. J Am Chem Soc 126:5309–5317
68. Takaluoma EM, Takaluoma TT, Oilunkaniemi R, Laitinen RS (2015) Structure and bonding in Bis(1-naphthyl) diselenide and bis{[2-(N , N -dimethylamino)methyl]phenyl} tetraselenide, and their brominated derivatives. Z Anorg Allg Chem 641:772–779
69. Zhu C, Yoshimura A, Ji L, Wei Y, Nemykin VN, Zhdankin VV (2012) Design, preparation, X-ray crystal structure, and reactivity of O-alkoxyphenyliodonium bis(methoxycarbonyl) methanide, a highly soluble carbene precursor. Org Lett 14:3170–3173
70. Patra A, Wijsboom YH, Leitus G, Bendikov M (2009) Synthesis, structure, and electropolymerization of 3,4-dimethoxytellurophene: comparison with selenium analogue. Org Lett 11:1487–1490
71. Mikherdov AS, Kinzhalov MA, Novikov AS, Boyarskiy VP, Boyarskaya IA, Dar'in DV, Starova GL, Kukushkin VY (2016) Difference in energy between two distinct types of chalcogen bonds drives regioisomerization of binuclear (diaminocarbene)Pd II complexes. J Am Chem Soc 138:14129–14137

72. Bader RFW (1991) A quantum theory of molecular structure and its applications. Chem Rev 91:893–928
73. Ho PC, Szydlowski P, Sinclair J, Elder PJW, Kübel J, Gendy C, Lee LM, Jenkins H, Britten JF, Morim DR, Vargas-Baca I (2016) Supramolecular macrocycles reversibly assembled by Te…O chalcogen bonding. Nat Commun 7:11299
74. Knowles RR, Jacobsen EN (2010) Attractive noncovalent interactions in asymmetric catalysis: links between enzymes and small molecule catalysts. Proc Natl Acad Sci U S A 107:20678–20685
75. Davis HJ, Phipps RJ (2017) Harnessing non-covalent interactions to exert control over regioselectivity and site-selectivity in catalytic reactions. Chem Sci 8:864–877
76. Wheeler SE, Seguin TJ, Guan Y, Doney AC (2016) Noncovalent interactions in organocatalysis and the prospect of computational catalyst design. Acc Chem Res 49:1061–1069
77. Cavallo G, Metrangolo P, Milani R, Pilati T, Priimagi A, Resnati G, Terraneo G (2016) The halogen bond. Chem Rev 116:2478–2601
78. Benz S, López-Andarias J, Mareda J, Sakai N, Matile S (2017) Catalysis with chalcogen bonds. Angew Chem Int Ed 56:812–815
79. Benz S, Mareda J, Besnard C, Sakai N, Matile S (2017) Catalysis with chalcogen bonds: neutral benzodiselenazole scaffolds with high-precision selenium donors of variable strength. Chem Sci 8:8164–8169
80. Wonner P, Vogel L, Düser M, Gomes L, Kniep F, Mallick B, Werz DB, Huber SM (2017) Carbon-halogen bond activation by selenium-based chalcogen bonding. Angew Chem Int Ed 56:12009–12012
81. Fukata Y, Asano K, Matsubara S (2015) Facile net cycloaddition approach to optically active 1,5-benzothiazepines. J Am Chem Soc 137:5320–5323
82. Suzuki T, Fujii H, Yamashita Y, Kabuto C, Tanaka S, Harasawa M, Mukai T, Miyashi T (1992) Clathrate formation and molecular recognition by novel chalcogen-cyano interactions in tetracyanoquinodimethanes fused with thiadiazole and selenadiazole rings. J Am Chem Soc 114:3034–3043
83. Zhao H, Gabbaï FP (2010) A bidentate Lewis acid with a telluronium ion as an anion-binding site. Nat Chem 2:984–990
84. Garrett GE, Carrera EI, Seferos DS, Taylor MS (2016) Anion recognition by a bidentate chalcogen bond donor. Chem Commun 52:9881–9884
85. Benz S, Macchione M, Verolet Q, Mareda J, Sakai N, Matile S (2016) Anion transport with chalcogen bonds. J Am Chem Soc 138:9093–9096
86. Dutton JL, Martin CD, Sgro MJ, Jones ND, Ragogna PJ (2009) Synthesis of N,C bound sulfur, selenium, and tellurium heterocycles via the reaction of chalcogen halides with –CH 3 substituted diazabutadiene ligands. Inorg Chem 48:3239–3247

Index

© Springer International Publishing AG, part of Springer Nature 2018
E. J. Lenardão et al., *New Frontiers in Organoselenium Compounds*,
https://doi.org/10.1007/978-3-319-92405-2

The manufacturer's authorised representative in the EU is Springer Nature Customer Service Centre GmbH, Europaplatz 3, 69115 Heidelberg, Germany. If you have any concerns regarding our products, please contact ProductSafety@springernature.com

Printed and bound by CPI Group (UK) Ltd, Croydon, CR0 4YY
05/10/2026
02228139-0001